Penguin Education
Penguin Library of Physical Sciences

Matter and Motion
N. Feather, F.R.S.

Advisory Editor
V. S. Griffiths

General Editors
Physics: N. Feather, F.R.S.
Physical Chemistry: W. H. Lee
Inorganic Chemistry: A. K. Holliday
Organic Chemistry: G. H. Williams

Matter and Motion

N. Feather, F.R.S.

Penguin Books

Penguin Books Ltd, Harmondsworth,
Middlesex, England
Penguin Books Inc., 7110 Ambassador Road,
Baltimore, Md 21207, U.S.A.
Penguin Books Australia Ltd, Ringwood,
Victoria, Australia

First published 1970
Copyright © N. Feather, 1970

Made and printed in Great Britain by
R. & R. Clark Ltd, Edinburgh
Set in Monotype Times New Roman

Contents

Editorial Foreword

For many years, now, the teaching of physics at the first-degree level has posed a problem of organization and selection of material of ever-increasing difficulty. From the teacher's point of view, to pay scant attention to the groundwork is patently to court disaster; from the student's, to be denied the excitement of a journey to the frontiers of knowledge is to be denied his birthright. The remedy is not easy to come by. Certainly, the physics section of the Penguin Library of Physical Sciences does not claim to provide any ready-made solution of the problem. What it is designed to do, instead, is to bring together a collection of compact texts, written by teachers of wide experience, around which undergraduate courses of a 'modern', even of an adventurous, character may be built.

The texts are organized generally at three levels of treatment, corresponding to the three years of an honours curriculum, but there is nothing sacrosanct in this classification. Very probably, most teachers will regard all the first-year topics as obligatory in any course, but, in respect of the others, many patterns of interweaving may commend themselves, and prove equally valid in practice. The list of projected third-year titles is necessarily the longest of the three, and the invitation to discriminating choice is wider, but even here care has been taken to avoid, as far as possible, the post-graduate monograph. The series as a whole (some five first-year, six second-year and fourteen third-year titles) is directed primarily to the undergraduate; it is designed to help the teacher to resist the temptation to overload his course, either with the outmoded legacies of the nineteenth century, or with the more speculative digressions of the twentieth. It is expository, only: it does not attempt to provide either student or teacher with exercises for his tutorial classes, or with mass-produced questions for examinations. Important as this provision may be, responsibility for it must surely lie ultimately with the teacher: he alone knows the precise needs of his students – as they change from year to year.

Within the broad framework of the series, individual authors have rightly regarded themselves as free to adopt a personal approach to the choice and presentation of subject matter. To impose a rigid conformity on a writer is to dull the impact of the written word. This general licence has been extended even to the matter of units. There is much to be said, in theory, in favour of a single system of units of measurement – and it has not been overlooked that national policy in advanced countries is moving rapidly towards uniformity under the *Système International* (S.I. units) – but fluency in the use of many systems is not to be despised: indeed, its acquisition may further, rather than retard, the physicist's education.

A general editor's foreword, almost by definition, is first written when the series for which he is responsible is more nearly complete in his imagination (or the publisher's) than as a row of books on his bookshelf. As these words are penned, that is the nature of the relevant situation: hope has inspired the present tense, in what has just been written, when the future would have been the more realistic. Optimism is the one attitude that a general editor must never disown!

January 1968   N. FEATHER

Preface

Those who intend to base an undergraduate course in physics on the volumes of the Penguin Library of Physical Sciences will probably wish to assign this volume for study in the first term of that course. It has been written with precisely that use in mind.

Nearly a century ago, James Clerk Maxwell contributed a slim volume entitled *Matter and Motion* to the series of Manuals of Elementary Science published by the Society for Promoting Christian Knowledge. He wrote in his Preface (1876), 'To become acquainted with these fundamental ideas [of configuration, motion and force], ... must be the foundation of the training of the student of Physical Science.' A similar statement is equally true today – though the inter-relations of our 'fundamental ideas' may have changed.

It is with some presumption on the part of the author that Maxwell's title has been appropriated for this book. Peter Guthrie Tait, writing an account of Maxwell's life and work for the *Encyclopaedia Britannica*, said of the original *Matter and Motion*, '[it] is ... worthy of the most attentive perusal not of students alone but of the very foremost scientific men.' Assuredly, no such quality attaches to the present volume, but the author modestly hopes that some students, having read it once in the first year of their university studies, will not find it unprofitable to read it again before their final examinations. Or they might even read it for pleasure thereafter!

Chapter 1
Introduction: the Scope and Method of Physics

Today, physics is reckoned, simply, as one of the sciences, though there is considerable justice in the claim that it is the basic science. In origin, the word is the Greek word for Nature – for the cosmos. Now it has a humbler connotation. Yet the physicist of the twentieth century cannot altogether disown the past: the physics and the cosmology of the classical period of Greek culture are part of his inheritance – he would be the poorer if it were not so.

One of the earliest of the Greek physicist-philosophers was Anaxagoras (c. 500–428 B.C.). As a young man he had stated clearly what he saw to be his destiny: 'I was born,' he said, 'that I might contemplate the works of nature.' In his old age they cast him out of Athens for the impiety of his views – for his enemies had roused the common people against him. Contemplation of nature epitomizes the Greek view. Plato (427–347 B.C.), who was born in the city which had denounced Anaxagoras, about the time of his exile, saw the world as a work of art. The cold, dispassionate beauty of it was for man's intellect to encompass: to appreciate that beauty, to see something of the compulsion and the freedom of the Artist as he fashioned it, in the heavens and on earth, was man's privilege. There was no thought of material reward, or of political power ensuing; only the adventure of the mind, ranging the universe and knowing it to be good.

Physics began, then, two and a half millennia ago, as the contemplation of nature. In that process, at the beginning, there was more of inspiration than of discipline. The strict discipline of logic was there, as the Greeks saw it, but there was not the discipline of experiment. That is the crucial matter, as we shall discover. There was inspiration in plenty: general ideas were formulated, concepts were framed, and human reason ran riot in a welter of possibilities. Aristotle (384–322 B.C.) wrote eight books under the general title *Physics*, an imposing monument to the human intellect – but that particular monument is in ruins, just because the foundations of experiment were not there. Indeed, nothing but ruins remains of the physics of the Greeks – nothing, save the firm belief among physicists that the world is ordered and able to be understood, the belief in universal categories of description, in universal causes or forces, call them what you will: the belief in permanence underlying change. Surely this is enough. The twentieth-century physicist cannot disown the past, for these beliefs are basic for his whole endeavour. In the opening sentence of *Matter and Motion* (1876), James Clerk Maxwell (1831–79) wrote, 'Physical Science is that department of knowledge which relates to the order of nature.' In that he was merely echoing the Greek philosophers; his definition may be accepted today, without significant change.

Modern physics began to take shape with the experiments of Galileo (1564–1642) and Newton (1642–1727). The philosophy of the Greeks had been a speculative philosophy; these men called themselves natural philosophers (even today, in many of the older universities, our subject goes by the name of natural philosophy, so maintaining unbroken the tradition of four centuries' study). Newton summarized the new procedure: 'In this philosophy particular propositions are inferred from the phenomena, and afterwards rendered general by induction.' In another place he wrote, 'the arguing from experiments and observations by induction ... is the best way of arguing which the nature of things admits of ... if no exception occur from phenomena, the conclusions may be pronounced generally. But if at any time afterwards any exception shall occur from experiments, it may then begin to be pronounced with such exceptions as occur.' This may not be very profound metaphysics, but it is sound common sense.

Modern physics, indeed, is ultimately based on simple observations and common sense. In this aspect its history stretches back long before the time of Galileo, or of the philosophers of Greece, to the early civilizations of Egypt and Mesopotamia. There are simple measurements of the things of the world that men have come to make in the conduct of trade, in the arts of peace and the business of war, which have developed gradually over the centuries, which have not required elaborate equipment or strictly logical definition of procedure. Likewise, out of everyday experience have arisen the notions of cause and effect, and the feeling of what is relevant and what is irrelevant, in a given situation and in respect of the particular interest of the time. The modern physicist has merely refined these procedures, imposing discipline on the intuitive notions inherited from his ancestors, greatly increasing the range of his own awareness of his physical environment. He has done this by confining his attention to those aspects of the world which can be submitted to measurement, for his is a quantitative science. He has nothing to say on the subject of value, in human estimation or divine.

As we have just implied, the story of the development of physics is twofold. It concerns the sharpening of our senses and the imposition of discipline on our intuitions. We survey the world with a human yardstick. Things which are not too large or not too small are open to our immediate inspection; they have been familiar to us from the beginning. Objects immensely larger and immensely smaller than these have been brought within the range of the familiar only more recently: over the last three centuries and a half the telescope and the microscope have greatly sharpened our sense of sight. Our sense of touch has been sharpened, in one of its sensitivities, by the thermometer, our sense of hearing by the microphone. Through the use of the ammeter and the voltmeter we have in effect acquired senses that we did not previously possess – natural man has no specifically electrical sense organ. All these aids to sense have enormously expedited progress in our subject; in many respects they have been indispensable. However, the discipline that has been imposed on our intuition has been even more important. It has been fundamental for success. Here, then, is the core of the matter: the interplay of speculation and proof, the sifting of intuitive notions by the discipline of experiment – the receptive mind continually probing, the perceptive mind continually imagining new possibilities of connexion between phenomena, a coherent picture of the world slowly coming into focus in the mind's eye.

Newton, as we have seen, found it necessary to distinguish between 'observation' and 'experiment'. In principle this is an important distinction, though very few physicists today can ever be involved, professionally, in pure observation. Pure observation almost precludes measurement: it requires concentration of critical attention on a particular sequence of 'natural' events which the 'observer' does nothing consciously to influence. He isolates these events by narrowing the field of his perception to hold them in attention, that is all. His record of them is necessarily an interpretative account, the language that he uses for the purpose being conditioned by the framework of general ideas already formed in his mind. By contrast, an experiment involves conscious intervention by the experimenter. He attempts to isolate phenomena for study by positive action, rather than by mental concentration, by devising 'artificial' situations in which the environment is well defined – and, in general, variable at will. Perhaps even more than the passive observer, the experimenter is limited in his approach by the 'theoretical' ideas which together constitute his view of the world. He devises his experimental arrangement to search for an effect which is compatible with notions already entertained. If his theoretical ideas are mistaken, it is unlikely that he will find the effect for which he is looking. If they are not too far removed from 'the truth', he may still have achieved something in understanding – a negative result may occasionally be as useful as any other – but the ill-conceived experiment which leads nowhere in particular is not unknown, in the long history of our subject.

In relation to both experiment and observation, we have stressed the notion of the isolation of phenomena. This is an essential aspect of the methodology of physics, and of science in general. Holding a belief in the orderliness of nature, the physicist is prepared to find that every event in the external world has some relation with – is partially determined by, or partially determines – every other event. Yet, in order to make significant progress towards understanding, he is forced to act in disregard of this belief. When he performs an experiment, almost inevitably he adopts the fiction that he is working with an ideally isolated system. This may be a good approximation to the truth, or it may not: it is no more than an approximation, in any case, whether good or bad. Conceivably, the world could have been other than it is; conceivably, it could never have been possible for a man-contrived situation to approximate closely to the conditions of isolation that the physicist requires. In fact, the procedure is found 'to work', but the experimenter should never lose sight of the basic assumption which it entails.

13   The Scope and Method of Physics

It is popularly said that the physicist discovers the laws of nature. Orderliness implies the rule of law: it is the physicist's business to unravel the complexities of phenomena, so it is said, to lay bare the underlying regularities and expose the natural law. All this presupposes his ability to communicate. Though he starts from the standpoint of common experience, the physicist soon finds that the language of everyday life is inadequate for his purposes. He must give precision to the common-sense categories of thought which he is able to use, and he must develop others of his own. He must formulate new concepts and precise definitions – and he must ever be wary of their proliferation. All the time he must remember that his subject advances, in the last resort, only by measurements made in experimental situations. His concepts, therefore, should have relevance to experiment: indeed, they should in general issue from the consideration of experiments already made, forced on his attention by the emergence of an aspect of regularity not previously recognized.

The physicist discovers the laws of nature – and ultimately expresses them in mathematical form. Ideally, the measure of his success – paradox though it may appear – is the smallness of the number of independent laws of nature which he claims to have discovered. Ideally, if the cosmos is 'rightly ordered' in a grand design, that design should be 'woven in a piece throughout'. Today, no physicist would be foolish enough to suppose that his knowledge of the physical world is all but complete, yet there is very wide knowledge – and critical evaluation shows that there are very few fundamental laws. The textbooks are full of equations, but very few, on examination, enshrine any natural law. Most of them stem directly from definition, useful, indeed necessary, for the organization and exploitation of knowledge, and for the conduct of experiment, but very few make any independent statement about the natural world that has not been written into them in the course of their development. In the last analysis, the fundamental laws of nature are those which describe the 'forces' which determine the structure and behaviour of bodies in general, that give the actual universe its character of stability and change (which might, conceivably, have been other than it is), and these, as it appears, are few in number. Early in the nineteenth century, physicists recognized as of different kinds the forces of electricity, of magnetism and of gravitation – and the chemists might have claimed that the force of chemical affinity is different again. Nowadays, the situation is changed and simplified. As far as tangible matter is concerned, there is gravitational force and electrical force; the phenomena of magnetism are recognized as arising from the effects of electricity in motion, the phenomena of chemical affinity are electrical in origin, also. At this macroscopic level, just two types of force suffice for the physicist's description of the world. Of course, that is not the whole story: the nuclear physicist of the twentieth century has opened up a new world of the intangible which is full of detail and surprise. At the present time he would probably insist that his understanding of this sub-microscopic world requires two, or possibly three, 'new' types of force, in addition to the others. In this he may be right, or time may bring a simplification here, also.

It is immaterial for our purpose: throughout the whole world – and it must be stressed that the world of nature is one and indivisible – we recognize at the most five types of force, distinct and individual: our fundamental laws are few in number, indeed.

Further reading

R. E. Peierls, *The Laws of Nature*, Allen & Unwin, 1955.
V. F. Weisskopf, *Knowledge and Wonder*, Doubleday, 1962.
A. N. Whitehead, *The Concept of Nature*, Cambridge University Press, 1919.

Chapter 2
Space and Geometry

2.1 Historical

Geometry is by far the oldest of the practical sciences. The name which its early practitioners gave to it indicates their general aim: 'earth measurement'. As providing the rules for such measurement, practical geometry was developed by the surveyors of Egypt and Mesopotamia through many centuries – and practised for more than two thousand years – before the rise of the civilization of Greece. With the Greeks, notably with Thales of Miletus (c. 624–547 B.C.), Pythagoras (c. 572–497 B.C.) and Eudoxus (c. 408–355 B.C.), this practical science was transformed into a formal, logic-bound system, the first branch of mathematics. Today, the word generally bears that interpretation – moreover, there are now many geometries, differing one from another according to the axioms from which they are developed, in the logical process of thought.

Because the formal geometry of the Greeks developed out of the practical science of earth measurement, it is inevitable that, even at this distance in time, it should have some relevance to the real world, though its basis is axiomatic. Indeed, the Greek philosophers thought of it as exhibiting the 'perfection' of actual space – empty or occupied by real bodies – unsullied by the inconsistencies which derive from imperfect measurement. To a large extent we must admit this claim as a reasonable one. When the civilization of Greece had passed its zenith, in the third century B.C., the work of the great geometers was systematized by Euclid. Thereafter, for more than two thousand years, it was believed that the three-dimensional space of Euclid's *Elements* provided an exact means of representation of the extensive attributes of bodies and of the configurations in which they are found in the world. It satisfied alike the intuitive notions and accumulated experience of common men and the mathematical requirements of the physicist. The belief that Euclidean space is the exact formal counterpart of the space of experience (in which events happen in a uniquely determinable sequence in time) proved to be no embarrassment to the physicist until quite recently: until we come to the last chapter of this book we shall adopt it without question.

In its purely formal aspect, Euclidean space is featureless – it is an infinite void, without 'landmarks' of any kind. We owe to René Descartes (1596–1650) the development of the system of coordinate geometry whereby the position of any point in space is specified in terms of its perpendicular distances from three arbitrarily situated, mutually perpendicular planes. These three distances are the *Cartesian coordinates* of the point (usually denoted by the composite symbol (x, y, z)), and the three lines in which the reference planes intersect in pairs are the *Cartesian (rectangular) axes* concerned. The common point, through which all three axes pass, is the *origin of coordinates*. Used in relation to a physical system, a set of Cartesian axes (or any other appropriate set of axes) provides a 'frame of reference' for the discussion of the spatial characteristics of the events occurring within that system.

Euclidean space is featureless – and impersonal. To establish a frame of reference involves the choice of an origin, and of the directions of the axes. It is virtually impossible to give meaning to such a procedure of choice unless it refers to non-empty space. In practice the origin will be fixed in some physical object of interest to the experimenter. Only a local frame of reference will generally be required. We started from the concept of infinite Euclidean space: very quickly we have 'come down to earth' – even the door of the laboratory has closed behind us. At this stage the important comment to make is that physicists in their individual laboratories, over a period of three hundred years, effectively set up their personal frames of reference against which they made measurements of the spatial attributes of the systems which they investigated – and, leaving aside uncovenanted error, found themselves in agreement in all essential particulars, when the results of their investigations were compounded. Clearly, within the range of variation of circumstance involved, the appearance of the physical phenomena concerned did not in any way depend on the choice of a reference frame.

Physicists, of course, are not exclusively concerned with laboratory experiments. In a later chapter we shall be dealing in some detail with the phenomena of gravitation. At least in the earlier stages of the investigation of these phenomena, physicists had to rely on observation rather than controlled experiment – observation of large-scale 'astronomical' events. For this purpose frames of reference were required against which the apparent motions of the celestial bodies could be determined. As we shall see, choice of reference frame in these circumstances is not a matter of indifference. Against a frame of reference having its origin in the astronomer's observatory the apparent motions of the planets are of baffling complexity; against a frame of reference having its origin at the centre of the sun they take on an over-all simplicity which cannot but be regarded as significant for understanding. Essentially it was the achievement of Copernicus (1473–1543) to make this transfer of origin, in thought, and draw the attention of his contemporaries to the great simplification thereby achieved.

We have just said that it is virtually impossible to give meaning to the procedure of choice of reference frame unless it refers to non-empty space. This statement is equally valid whether it concerns a 'local' frame of reference for a laboratory experiment, or an 'infinite' frame of reference for the astronomer's observations: once we decide to 'fix' the origin of our Cartesian coordinates without any reference whatever to the positions of material bodies, one region of space is exactly like any other – and the notion of choice becomes meaningless. The notion of 'absolute rest in space' (which might have been exemplified by the condition of the origin of a uniquely identified set of rectangular axes – if such a set could have been identified in 'absolute space') is meaningless by the same token. Gottfried Leibniz (1646–1716) was perhaps the first to insist on the inevitability of this conclusion. He wrote, briefly, 'Space is the abstract of all relations of co-existence.' Much later, and with more humour, Maxwell wrote (*Matter and Motion*, p. 20), 'Any one, however, who will try to imagine the state of a mind conscious of knowing the absolute position of a point will ever after be content with our relative knowledge.' 'Absolute space' and 'absolute rest' are misconceived notions, for the physicist or for anyone else.

2.2 Configurations of material particles

We shall be concerned generally in this book with the motions of material bodies. Motion is progressive change of position in space: it involves the idea of time implicitly. Later, we shall have to consider the idea of time in its own right; for the present we are more concerned to examine the spatial relations of a system composed of real bodies 'at an instant in time'. Clearly, this notion of the state of a system at an instant is an abstraction from reality – even in 'snapshot' photography the duration of exposure is finite (and its proper choice is important). Moreover, we are unlikely to be interested in the instantaneous configuration of any system unless we can, at a later time, or times, identify the constituent bodies of the system, and so trace their history. Physics is concerned with the discovery of what is permanent in the changing appearance of the world.

We have already drawn attention to one fiction which the physicist is compelled to adopt in relation to his experimental arrangements – the fiction of the isolated system; likewise, as far as he is able, he adopts the fiction that the material bodies with which he is concerned possess enduring identity. He knows that this, also, is only an approximation to the truth: lumps of metal are slowly abraded 'by wear and tear' (old coins are withdrawn from circulation when the process of abrasion exceeds an acceptable limit), and samples of liquid evaporate more or less slowly, according to circumstances. However, the physicist of today believes that all bodies are made up of atoms: charged with the fiction of permanence which he is apt to attribute to gross bodies, he is able to shift his ground and claim that the primordial atoms are permanent. Yet he knows, in these latter days, that that assertion, again, is strictly untenable. So he shifts his ground once more, and searches for permanence within the atom.

Scientific atomism became explicit with Dalton's *New System of Chemical Philosophy* in 1808, but the general idea of atoms had been accepted by natural philosophers for two centuries before that, indeed it can be traced back to Democritus in the fifth century B.C. It was an ingredient in the natural philosophy of Newton as set out in the *Principia* (1687). The system of mechanics which Newton established in that remarkable book has traditionally been presented in two parts: the first part deals with systems of discrete particles, the second treats of the motions of gross bodies. Formally, the second part is a generalization of the first: gross bodies were regarded as particle systems having permanent coherence. When we come to consider the matter in detail, we shall discover (see p. 138) that there is something wholly artificial in the traditional development of the Newtonian mechanics of particles – we should rightly be dealing with real bodies (however small) from the outset – but it is an educative exercise, nonetheless, and it has much to contribute to our ultimate understanding. Maxwell's definition of a material particle (*Matter and Motion*, p. 11), with its following gloss, is highly instructive – and it points directly to the artificiality that we have mentioned. Maxwell wrote:

A body so small that, *for the purposes of our investigation*, the distances between its different parts may be neglected, is called a material particle. Thus in certain astronomical investigations the planets, and even the sun, may be regarded each as a material particle.... But we cannot treat them as material particles when we investigate their rotation. Even an atom, when we consider it as capable of rotation, must be regarded as consisting of many material particles.

In the formal treatment of the *Principia*, Newton did not normally refer to the constituent particles of bodies as 'atoms': we shall, however, do this, for the atom concept is now well defined. On the other hand, we shall not speak of atoms when we are dealing with 'systems of Newtonian particles'–in that connexion we shall accept Maxwell's definition, noting the caution which he attached to it.

In this section, as we have stated, we are not concerned with motion, either of translation or rotation, but merely with configuration at an instant in time. Having clarified our ideas concerning the constitution of the bodies which make up the systems with which we shall be concerned, we are left merely with matters of definition. We say that the *configuration* of an 'isolated' material system is the assemblage of the relative positions of all its parts. More specifically, we may say that the configuration of a material system at an instant in time is given in terms of the positions of all the atoms constituting the bodies of the system at that instant, these positions being specified in relation to a frame of reference chosen with respect to the system. Clearly, for purposes of this definition it is a matter of no consequence whether the system is a system of 'gross bodies' or of 'particles', in the Newtonian sense.

In thought, as in this particular definition, we imagine frames of reference set up in the space of experience. This thought process is natural, and indeed necessary, if our considerations are to fulfil their purpose of having direct relevance to the real world. But the very naturalness of the process obscures an important distinction which we should make explicit at the first opportunity, for it is basic to our whole procedure. We commit our thoughts to paper, and carry forward our arguments through the use of mathematical symbols and diagrams. The frame of reference which we set up, in thought, in the space of experience is represented formally by a conventional geometrical figure showing a set of Cartesian axes on a two-dimensional surface (or it could be represented by a three-dimensional model, though that is more cumbersome). It would be foolish to suggest that, immersed in a particular problem, we should all the time be reminding ourselves that our diagrams are 'only diagrams and not the real thing' – fluency in the use of such diagrams is of first importance to the physicist – but we should at least, from time to time, recall this fact, for it is one of the pitfalls of our subject to regard our symbols and diagrams as real, when they are no more than human artefacts, purely conventional representations of the measurable attributes of the world. We have introduced this consideration 'at the first opportunity', in relation to diagrams of configuration. In this connexion there is a very obvious relationship between those measurable attributes of the physical systems with which we are dealing and the features of the diagrams that we use for their representation. Later, we shall be using diagrams to represent other measurable attributes of systems of particles – velocities, accelerations, forces, and the like. Obviously, our note of caution is not entirely pedantic.

2.3   Representation by vectors

The position of a point with respect to a set of Cartesian axes is given by three measures of length (p. 17). It may equally well be specified by one length and two angles. Suppose that P (Figure 1) represents the point whose Cartesian coordinates with respect to the rectangular axes $X'OX$, $Y'OY$ and $Z'OZ$ are (x, y, z). Let PN be perpendicular from P on the plane XOY, and NA and NB be perpendicular, in this plane, on OX and OY, respectively. Then, on the scale of the diagram, $OA = x$, $OB = y$, $NP = z$. Our contention is that, if $OP = r$, $\angle ZOP = \theta$, $\angle AON = \phi$, then the *polar* coordinates (r, θ, ϕ) specify the position of P as effectively as do the Cartesian coordinates (x, y, z). Obviously, from the figure,

$OA = ON \cos \phi = OP \sin \theta \cos \phi,$
$OB = ON \sin \phi = OP \sin \theta \sin \phi,$
$NP = OP \cos \theta,$

that is
$$\left. \begin{array}{l} x = r \sin \theta \cos \phi, \\ y = r \sin \theta \sin \phi, \\ z = r \cos \theta, \end{array} \right\} \qquad \qquad \textbf{2.1}$$

or $r^2 = x^2 + y^2 + z^2 \qquad \cos \theta = \dfrac{z}{r} \qquad \tan \phi = \dfrac{y}{x}. \qquad \textbf{2.2}$

Equations **2.2**, establishing the one-to-one correspondence between the two sets of coordinates (on the basis of the convention that r is always a positive quantity), establish our contention explicitly.

The specification of the position of a point in terms of its polar coordinates

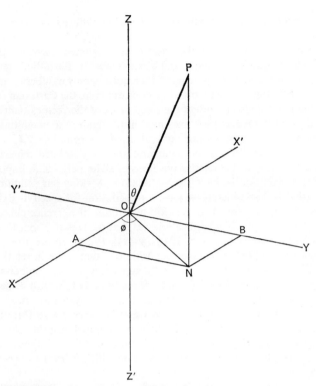

Figure 1

(r, θ, ϕ) presupposes the choice of an origin and of two reference directions (OZ and OX in the figure). It is then given by the length and direction of the line drawn from the origin to the point. Such a directed line $\overrightarrow{OP}$ is the simplest example of a *vector*. In this connexion the vector is strictly the 'displacement' which would transfer (or 'carry') a point at the origin to the position of the specified point (r, θ, ϕ). As so described, displacement is a *vector quantity*.

The clear notion of a vector, as it is accepted today, was slow to emerge from the considerations of the mathematicians. John Wallis (1616–1703) had the rudiments of the idea in 1685 but it did not reach full development for another century and a half. Eventually it was given precision through the work of William Rowan Hamilton (1805–65) and Hermann Grassmann (1809–1877). Now we have a fully developed vector algebra with its own rules and formalism. We shall not use this formalism systematically in the present book, but we shall recur to it from time to time as occasion serves. Here we need only say that it is possible to use single symbols, consistently, to represent the full information that is necessary to characterize vector quantities of any kind (information, that is, concerning magnitude and direction). In order to identify these symbols, conventionally, they are set in boldface type. Thus, in our present example, the vector quantity 'displacement', which relates the position of the point P (Figure 1) to the origin O, is designated **r**, whereas the magnitude of this quantity (direction apart) is represented by the italic letter r (as is normal custom when physical magnitudes requiring only numbers – and units – for their specification are involved). In this example, the Cartesian coordinates of the point P also represent the *components* of the vector quantity **r** in the directions of the rectangular axes used in the figure: the magnitudes of these components are given, in terms of r, θ and ϕ, by equations **2.1**.

In the last section we defined the configuration of an isolated material system as the assemblage of the relative positions of all its parts. The important word in this definition is the adjective 'relative'. Making our definition more specific (that is, less general) we gave a prescription for specifying the positions of all the parts (atoms) 'in relation to a frame of reference chosen with respect to the system'. The implication was that the 'relative' positions of the parts would thereby be given. If (x_k, y_k, z_k), (x_l, y_l, z_l) are the coordinate specifications of the instantaneous positions of two atoms in relation to our chosen frame of reference, we tacitly suppose that the number-set $(x_l - x_k, y_l - y_k, z_l - z_k)$ specifies the position of the second atom with respect to the first. Indeed, this set of coordinates would necessarily characterize the position of the second atom with respect to Cartesian axes drawn through the position of the first atom as origin, in directions parallel, respectively, to those of the original reference frame. There is nothing doubtful about our supposition; in Euclidean geometry the proposition which it involves is self-evidently true.

In respect of configuration, which is our sole concern at the moment, the considerations of the last paragraph may appear little more than trivial. As we have defined it, the configuration of a system is an internal attribute of the system at an instant of time. The over-all character and detailed features of the configuration are clearly independent of the Cartesian frame of reference used for its description. They are equally independent of the choice of origin and reference directions for the description of the configuration in terms of the polar coordinates of its constituent atoms, that is of the choice of origin for its vectorial description. Our object is to construct a map of the spatial disposition of the constituent atoms of the system. If our map, in the end, is a map (diagram) constructed in terms of displacement vectors, we are in no way interested in the individual vectors drawn from our arbitrary origin to the points representing the positions of the various atoms in actual space, but only in the displacement vectors relating every such point in the diagram to every other point. In relation to the original construction involving vectors drawn from our chosen origin, these particular vectors may be characterized as 'difference vectors' – and this characterization leads us to consider, a little more formally, what we mean by addition and subtraction of vector displacements in this connexion.

We wish to give a meaning to the composite symbol $\mathbf{r}_1 + \mathbf{r}_2$, when $\mathbf{r}_1$ and $\mathbf{r}_2$

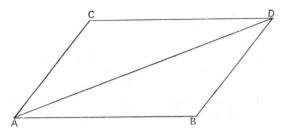

Figure 2

represent two displacement vectors. Consider a point A. In respect of this point the symbol r_1 describes a displacement, through a particular distance and in a particular direction, bringing A to the position B. Alternatively, r_2 describes the displacement of A to C, say. These alternative displacements can be represented in a two-dimensional diagram such as Figure 2, if the plane of the diagram is the plane through A defined by the directions of the two displacements. Because there is nothing in the meaning attached to r_1 or r_2, individually, to specify the point in space from which the corresponding displacement is to be effected, r_2 (for example) may be taken to refer either to the displacement $\overrightarrow{AC}$ of Figure 2, or to the displacement $\overrightarrow{BD}$, provided that BD is equal in length to AC and is parallel to it. This being the case, since the displacement $\overrightarrow{AB}$ is represented by r_1, the meaning that we should attach to $r_1 + r_2$ is that this composite symbol describes the result of the consecutive displacements $\overrightarrow{AB}$ and $\overrightarrow{BD}$ according to the figure. As a result of these consecutive displacements the point A is transferred to D. The same end-result could have been achieved by consecutive displacements $\overrightarrow{AC}$ and $\overrightarrow{CD}$. Consistently with our previous interpretation, therefore,

$$r_1 + r_2 = r_2 + r_1. \qquad 2.3$$

Equation 2.3 exemplifies the commutative law of addition in vector algebra.

Again, with reference to Figure 2, the displacement from A to D might have been along the straight line AD. In an obvious notation, then,

$$r_1 + r_2 = r, \qquad 2.4$$

r being the symbol representing the displacement $\overrightarrow{AD}$. Transposing equation 2.4, we have immediately

$$r_2 = r - r_1. \qquad 2.5$$

Here we have, formally, an equation of which the right-hand member is the difference of two vector displacements. We interpret it by saying (in relation

23 Representation by Vectors

to Figure 2) that the vector displacement which carries B to D is obtained by subtracting, from the vector displacement which carries A to D, that which carries A to B. Obviously this result is independent of the position of the 'origin' A. It is the basic result underlying our original assertion, 'If our map ... is constructed in terms of displacement vectors, we are in no way interested in the individual vectors drawn from an arbitrary origin to the points representing the positions of the atoms in actual space, but only in the displacement vectors relating every such point in the diagram to every other point.'

Before we leave this section we should add a gloss on the precise form of Figure 1. Conventionally, in respect of any set of rectangular axes, the directions $\overrightarrow{X'OX}$, $\overrightarrow{Y'OY}$, $\overrightarrow{Z'OZ}$ are taken as the positive directions of x, y and z. Similarly, in regard to rotation (or angular measure), the positive sense of rotation about $\overrightarrow{X'OX}$ ($\overrightarrow{Y'OY}$ or $\overrightarrow{Z'OZ}$) is taken to be that which would bring a straight line from coincidence with $Y'OY$ ($Z'OZ$ or $X'OX$) into coincidence with $Z'OZ$ ($X'OX$ or $Y'OY$) after rotation through one right angle. In relation to Figure 1 as drawn, these conventions associate positive rotation about any axis with positive translation along it in the same way as rotation and translation are associated (by mechanical constraint) when a 'right-handed' screw is turning in a fixed nut. For that reason a set of rectangular axes of the type shown in the figure is conventionally referred to as a set of right-handed axes. If we were to interchange the positions of the letters X and X' in Figure 1 (or Y and Y', or Z and Z') we should obtain a set of left-handed axes, according to the same usage. A little consideration will show that right-handed and left-handed rectangular axes stand to one another in the same relation as do an object and its image in a plane mirror – and, further, that this difference of 'handedness' (or 'cheirality') is the only non-trivial difference which two sets of rectangular axes can possibly exhibit.

In this book, when we make use of rectangular axes, we shall use right-handed axes exclusively.

2.4 Measurement of length and angle

Geometry began as a practical science (p. 16), but in the last two sections we have been engaged primarily in elucidating the more formal aspects of its use in the description of the configurations of particles, and of aggregates of particles, in space. For this purpose we assumed, tentatively, that the concepts of distance and angle were understood. In a textbook of physics, however, the problems of measurement cannot be overlooked: they are central to the main theme. Now, therefore, we redress the balance, discussing the measurement of length and angle systematically. Later, we shall discuss the measurement of other quantities, in a similar way, as the need arises.

The first necessity – whatever type of measurement we are concerned with –

is to choose a unit magnitude of the quantity involved. The criterion of suitability in this connexion is that our chosen unit shall be available for practical use as a 'measurable' magnitude characteristic of some physical system which can be supplied in 'faithful copies', on demand, to all the physical laboratories of the world (or constructed, at will, according to a precise recipe). This physical system provides the *standard* by which the unit magnitude in question is given material, as distinct from merely conceptual, expression.

The two quantities length (distance) and angle exemplify in a very direct way the alternative possibilities in relation to the setting up of standards that we have just distinguished. An accurately reproducible standard of angular measurement can be set up 'by recipe'; for a standard of length, on the other hand, the method of 'faithful copies' supplied by a central authority held sway in the civilized world for more than four thousand years. Only in the second half of the twentieth century, as we shall see, has 'specification by recipe' proved practicable in relation to this particular standard.

In Euclidean geometry a plane angle is enclosed between two intersecting straight lines, and its magnitude represents the amount of turning involved when either of the two lines rotates, about their point of mutual intersection, into coincidence with the other. Whatever the magnitude of the angle, a precise construction is available for its bisection. There is the limiting case when the two intersecting straight lines become congruent, being oppositely directed. It would be possible to take the angle 'between' the two lines in this situation as the unit of angular measurement – or to bisect this angle according to the recipe and take the so-called right angle as the unit. Indeed, in respect of this special case of congruent straight lines there is a unique construction (that of the equilateral triangle) by which an angle equal to one third of two right angles may be obtained. We might, therefore, take the common magnitude of the angles in an equilateral triangle as the unit.

Effectively, this last choice is the one that the master instrument makers of the eighteenth century made when the need for accurately divided circular scales for the telescopes of the period became urgent. Working from the angle which is one third of two right angles obtained in this way, they bisected it, continuing the operation six times in succession, until they obtained an angle of magnitude $\frac{2}{3}$ $(\frac{1}{2})^6$ of a right angle. In terms of the Euclidean recipe they thus constructed a master scale in which the unit was 1/96 of a right angle. Thereby a complete circular scale was ultimately subdivided into 384 equal parts – with an accuracy limited only by the instrument makers' skill in the use of the dividing compass and the sharpness of their engraving tools. This is what they did – to begin with – even though what the astronomers required was a circle divided into 360 equal *degrees*!

The right angle of 90 degrees (which we still employ) is a legacy of the distant past. It assigns 60 degrees to the angle of the equilateral triangle (or to the angle subtended at its centre by a side of a regular hexagon), and it is one in origin with the notation of sexagesimal fractions in arithmetic which derived from the Chaldeans. The arithmetic of sexagesimal fractions was adopted by the Greeks, and survived in general scientific use until it was replaced by the decimal notation of Henry Briggs (1556–1630) early in the seventeenth century. The Chaldean degree, with its sexagesimal parts, the minute and second of arc (1/60 and 1/3600 degree, respectively) escaped the zeal of the reformer, and survives today. The instrument makers of the eighteenth century had a precise recipe for the construction of a standard of angular measurement based on a right angle of 96 degrees: only through an accident of history were they unable to use that recipe to full advantage. In the end they had to do the best they could to interpolate 360 equally spaced divisions around the circumference of a circle which they had already subdivided into 384 equal parts. Their followers developed elaborate procedures for checking their final product for error, and gradually eliminating it. Although there was no precise recipe for their construction, in the end, 360° circles were produced for the great instruments which were well nigh perfect. Today we have almost lost sight of the fact that a unit of angular measurement is available that can be represented by a standard constructed according to recipe – nowadays circular scales of 360° are reproduced as 'faithful copies' of already existing scales of established quality, rather than graduated *de novo* using the constructions of Euclidean geometry.

The degree which is one ninetieth part of a right angle is the 'common' unit of angular measurement: as we have seen it is not in itself realizable by straightforward geometrical construction (though the right angle is). The 'mathematicians' unit' of angular measurement is the *radian* – the angle subtended at the centre of a circle by an arc equal in (curvilinear) length to the radius of the circle. More obviously than the degree, the radian is not realizable by simple construction: there is no formal way of drawing any curved line equal in length to a given straight line. Moreover, the ratio of the lengths of the circumference of a circle and its diameter is a surd – the most famous of all irrational numbers, denoted by π. A radian is $2/\pi$ right angles.

The distinction that we have made, between standards that can be set up by recipe and those which are fabricated as faithful copies of an ultimate standard, is generally drawn in other terms. We speak, alternatively, of 'natural' units and 'arbitrary' units: generally the standards representing the former can be specified by recipe, those representing the latter have to be made as copies of a unique primary standard. From this last point of view we may regard the radian as a natural unit of angular measurement (even though the recipe relating to its practical realization is unworkable), the Chaldean degree as an arbitrary unit.

The informed reader may well be critical of our using the problem of angular measurement as a first example of the wider problem of the choice of units and the establishment of standards of measurement generally. He will rightly insist (if he takes the critical view) that an angle is not a physical magnitude in the true sense: its measure is given by the ratio of two lengths (as the definition of the radian implies), that is all. Indeed, we should not speak of the 'magnitude' of an angle but only of its 'measure': it is a concession to human weakness to give a distinctive name to the 'natural unit' of angular measurement – angles are representable by pure numbers, directly. For that matter, we could do without angles altogether in relation to spatial configurations (though it would be most inconvenient): any information given in terms of polar coordinates can be given equally well in Cartesian form (p. 20). All this is unquestionably true, but it is equally true that in practice, in a multitude of situations, measurement of angle is the most significant measurement that we can make in the circumstances – and that no one uses scales divided directly in radians, and submultiples of radians, for this purpose. Another unit has been chosen for practical use – albeit long ago, and perhaps ill-advisedly – an arbitrary, rather than a natural, unit angle. We accept the validity of our reader's criticism, then – but we do not therefore admit the irrelevance of our example.

In respect of the measurement of length, there can be no doubt as to the status of the quantity involved. Distance is a concept involving the mutual relationship of two points only (three points are necessary for the specification of an angle); obviously it cannot be reduced to anything more fundamental (there is nothing to be said about a single 'point' in otherwise empty space). Length (or distance) is a physical magnitude in its own right. As we have already implied, throughout the period of recorded history standards of length measurement have been set up arbitrarily, and 'faithful copies' of 'the standard' have been distributed for use to those whose business it was to maintain uniformity of practice in trade or commerce in a given region. Thus, over two thousand years and more, in Egypt and Mesopotamia, the cubit was maintained with but little variation, the material standards in the form of metal bars being held in safe keeping by the priests. For nearly as long again, throughout the Christian era in western Europe, a similar practice prevailed, the civil governments rather than the priestly caste being the custodians of the standards. In the course of time different countries adopted different units – Britain the yard, France the metre, and so on. Until 1960 all these units were maintained in the form of material standards, legalized by appropriate resolution of the governments concerned. The 'imperial standard yard' was last defined by the Weights and Measures Act of 1878; the metre, first introduced as a national standard by decree of the French National Assembly of 1799, was redefined as the International Prototype Metre as a result of the Metric Convention of 1875.

Let us be perfectly clear what is required of a standard of length (as of any other physical magnitude, correspondingly). It must provide, as precisely as limited human experience can ensure, a fixed unit in terms of which the variations of lengths and distances in physical systems can be identified and measured. Here is the difficulty: a material standard is itself a physical system, and is subject to change like any other. Gradually, increased knowledge of the behaviour of solid materials in a variety of circumstances pointed the way towards a definition which would ensure reproducibility. The temperature of the material standard must be specified, for metal bars expand when heated – and the mode of support must be laid down, for metal bars sag under their own weight to an extent which is determined by the positions of the supporting members. Such a bar is longer when suspended vertically than it is when resting on a horizontal table at the same temperature. In view of these considerations, it is not surprising that the legal specifications of the standards become more involved in successive definitions. That, indeed, was the outcome in historical terms. In relation to fundamental physics the outcome can be differently assessed. We attempt to introduce an arbitrary unit of length and to give it material expression in terms of a practical standard. We find that we cannot do this satisfactorily unless we are able to measure other physical quantities – temperature, at least, and possibly others as well.

As we have twice implied (pp. 25, 27), material standards of length have now been abandoned as ultimate reference standards (see below). If they had been retained, at least one additional clause would have been necessary in any legal specification of a completely new standard. After the description identifying the metal bar and the graduations on it defining the unit length, and the specifications relating to temperature and the rest, there would have had to be the clause, 'this standard shall become the recognized standard of length at a date not earlier than one hundred years from the date of this Act'. This clause would have been necessary because of experience gained from the regular intercomparison of the imperial standard yard and its four 'parliamentary copies' over the last eighty years. Four of these five metal bars were fabricated in 1844–45; the fifth was produced in 1879 by a similar process of manufacture. Over the years this last bar has shortened in length in relation to the other four by about $2 \cdot 5 \times 10^{-4}$ inch. This process appears now (1967) to be approaching completion. Its observation brings to light an important phenomenon well known to engineers and metallurgists over a shorter time scale. 'Severe' treatment of any kind – rolling, extrusion, intense heating or rapid cooling – generally induces, in metallic objects, after-effects of longer or shorter duration. Experience with the imperial standard yard and its copies shows that, when an accuracy of one part in a million is in question, these after-effects may be significant over periods of tens of years at the least.

In 1960, at the eleventh General Conference of Weights and Measures (set up in perpetuity by the Metric Convention of 1875, to which we have already referred), the metre was re-defined as equal to '1 650 763·73 wavelengths in vacuum of the radiation corresponding to the transition between the levels $2p_{10}$ and $5d_5$ of the krypton-86 atom'. Three years later, the yard was redefined as 0·9144 metre precisely (intercomparison of the imperial standard yard and the international prototype metre in 1922 had yielded the result 1 yard = 0·914 398 41 metre). Clearly, the second of these definitions is more easily understood than the first – indeed, it is no more than a statement of equivalence, the legalization of a conversion factor for short-term use, until the metre should be adopted universally as the single internationally accepted unit of length measurement throughout the world.

In respect of the first definition, equally clearly, we cannot here explain all its terms in detail: we can only indicate their general significance. In the first place, the definition implies that the effective unit of length is now a 'wavelength' characteristic of an isolated atom – the atom of 'krypton-86'. It is therefore a natural, not an arbitrary unit. We have said that in such cases the standard is set up by 'by recipe'. In this case the recipe is designed to ensure that the wavelength of the radiation (visible light) emitted by the standard is that corresponding to emission by atoms of krypton-86 acting independently one of another. In its present form (for the recipe may be improved, even though the definition remain unaltered) it specifies that the light source shall be a discharge tube having a hot cathode and a central capillary section, that the discharge shall be viewed along the length of the capillary and through the anode side of the tube, that the current density of the discharge in the capillary section shall be 0.3 ± 0.1 A cm^{-2}, that the capillary section of the tube shall be immersed in a refrigerant bath maintained at a temperature within $1°$K of the triple point of nitrogen ($64°$K, or $-209°$C), and, finally, that the krypton-86 in the tube shall be of not less than 99 per cent purity and of sufficient quantity so that traces of solid krypton are to be seen in the tube at the working temperature.

So much, then, for the recipe specifying the standard – but it does not explain the physical principles underlying the definition of the unit. In order to appreciate these principles we have to recall that if a 'diffraction grating' is produced by ruling a large number of equally spaced parallel lines on a plane plate of glass (say 5000 lines per cm, across the plate) and if this grating is set up on the 'prism table' of an optical spectrometer, the lines of the grating being parallel to the axis of the instrument, then the grating 'analyses' an incident beam of parallel light into a 'spectrum' of coloured components. Indeed, in general, several similar spectra may be observed. If the incident light impinges normally on the grating, these spectra (first, second, ... order spectra) are symmetrically disposed about the direction of the incident beam. The simplest spectra are obtained with low-pressure gas discharge sources, particularly when monatomic gases are employed. These spectra then consist of no more than a few sharp 'lines' – and it is natural to suppose that the simplicity of the situation in such a case is due to the fact that in the low-pressure discharge individual atoms are acting as independent sources of radiation. For a particular spectral line (of a definite hue) it is found empirically that, in the symmetrical arrangement just described, the line appears, in its successive orders, at angular deviations θ_1, θ_2, θ_3, ... mutually related by the general expression

$\sin \theta_1 : \sin \theta_2 : \sin \theta_3 : \dots :: 1 : 2 : 3 : \dots.$

Clearly, in this expression we have to do with quantities which are pure numbers. Equally clearly, any expression representing the physical phenomenon with which we are concerned must involve at least one physical quantity relating to the grating and at least one relating to the light. In relation to the grating, its most obvious characteristic is a length – the common spacing of its rulings (in fact no other characteristic of the grating is significant for our present consideration); in relation to the light (of the well-defined spectral line) we therefore conclude that there is also involved a characteristic length. (A pure number may be obtained from the ratio of these two lengths.) It is this length that is the *wavelength* of the light. Effectively, measurements of angular deviation on the spectrometer enable us to compare the wavelength of the light with the grating spacing. As is established in textbooks of optics, the general grating formula is

$$n\lambda = d \sin \theta \qquad\qquad 2.6$$

(n being the order of the spectrum, λ the wavelength of the light, d the spacing of the grating and θ the angle of deviation concerned). If the grating had been ruled so that d was an exact (decimal) submultiple of the international prototype metre, our thought experiment would have resulted in a 'direct' comparison of the metre with the radiation wavelength that we had implicitly chosen as our natural unit of length.

The description that we have given indicates that it is possible to relate the wavelength of a visible radiation to the length of a measuring-rod standard: it must not be taken as doing more than that. In practice, much more sensitive methods of comparison are required than that which is possible using the grating spectrometer. Suffice here to say that such methods have been developed, and have now been in use for more than seventy years. The first such accurate comparison was made by Michelson and Benoit in 1894.

The 1960 definition of the metre specifies the atom of krypton-86 as the ultimate source of the standard radiation. In the twentieth century it has come to be recognized that the atoms of a pure chemical element are not, in general, all exactly alike: most elements consist of a mixture of *isotopic* atom species. The most important difference between the atoms of two *isotopes* of a given element is a difference in mass. To quite a high degree of accuracy (better than one part in a thousand in most cases) the masses of all atoms can be represented by integral numbers on a suitable scale. When krypton-86 is specified, it denotes the krypton isotope of mass number 86 on this scale.

We have already indicated that a monatomic gas is the most suitable filling gas for a discharge tube which is to be used for setting up a wavelength standard: krypton is such a gas. The requirement that a single isotopic constituent of this elementary gas be used ensures that the simplest possible spectrum shall be obtained. (The chemical properties of the various isotopic constituents of any element are the same 'as nearly as makes any difference': in a very high resolution spectrograph, on the other hand, it is found that each constituent produces its own spectrum; the various spectra are very nearly, but not exactly, the same; corresponding lines are very close together on the wavelength scale, so that it is practically impossible to work with an 'isolated' and 'sharp' line.) Now, the separation of a single isotopic constituent from a 'mixed' element is not a simple process. Krypton-86 is chosen because it is the heaviest isotope present in the natural gas. Any process of separation (which must depend on differences of mass) need only separate it from the lighter constituents: to have chosen any other isotope (except the lightest) would have imposed the demand that it be separated from other isotopic constituents both lighter and heavier than itself. Furthermore, there is a real advantage in specifying an isotope of even, rather than odd, mass number which we cannot elaborate here. Krypton-86 is an even-mass-numbered isotope. Altogether, this particular isotope of this inert gas has in high degree all the desirable characteristics which theoretical considerations of the problem of the standard wavelength source require.

Here we must leave the 1960 definition of the metre, merely remarking that the phrase 'radiation corresponding to the transition between the levels $2p_{10}$ and $5d_5$' is couched in the jargon of the atomic spectroscopist. For our purposes we need not attempt to understand it: the krypton spectrum, excited according to the recipe, is a spectrum of a few bright lines only. It will be enough if we substitute 'orange radiation' for the phrase in question. In that way we identify the radiation of interest by its colour (or hue). Colloquially, indeed, it is generally referred to as 'the krypton orange line'.

So the primary unit of length measurement is now designated as the wavelength of the orange radiation from a discharge tube containing krypton-86 and operated under well-defined conditions. For obvious reasons, this unit has not been given a new and distinctive name. It has been adopted solely because it can be realized in a standard with greater precision than any other unit of length measurement capable of definition at the present time. It is estimated that, in terms of practical reproducibility, the new unit is represented by the standard to one part in 10^8, or better. The reproducibility of the old metre, in terms of the international prototype (standard), was no better than about two parts in 10^7. Instead of giving the new unit a distinctive name, it is effectively designated as $6\cdot057\ 802\ 11 \times 10^{-7}$ metre by the formal definition (p. 28). Thereby it is recognized that, for practical purposes, 'the metre' remains the unit of measurement – and the international prototype metre, and its copies, retain their usefulness as secondary standards for many purposes. Indeed, the whole operation was designed to maintain the *status quo*, only to provide a more sure foundation for experimental procedure – and a more precise definition of the unit. We hopefully believe that the properties of individual atoms are not subject to slow change with the passage of time, as is the length of a metal bar fabricated in a rolling mill!

In the everyday work of the laboratory, measurements of length are made with the help of graduated scales, vernier callipers, screw gauges, and the like. Necessarily, these are measurements of low precision compared with the much more complicated measurements that have been made in the setting up of the wavelength standard and the designation of the new unit in terms of the metre. It is not our object, in this section, to discuss the principles underlying these laboratory procedures in detail. Suffice to say that in respect of length, rather than angle, there is no limitation in practical geometry to the feasibility of subdivision of a given magnitude. In general, an angle can only be bisected; a length can be divided into any number of equal parts by the appropriate construction. Once a length equal to the unit length (metre) has been marked off on a linear scale, there is a valid recipe for the subdivision of this length into whatever submultiples may be desired.

2.5 Area, volume and solid angle

Once the 'primary' unit of length measurement has been defined, 'derived' units of measurement of area and volume become available automatically. At least, provided we assume that the geometry of the space of physical experience is Euclidean, we may take over the familiar formulae of mensuration belonging to that system, and, inserting numerical measures of length in terms of the metre as unit, obtain measures of area and volume as so many 'square metres' or 'cubic metres', as the case may be. This being admitted for the regular figures, there can be no need in general to maintain a material standard representing the unit of area, or the unit of volume: the whole matter is determined unambiguously once the unit of length has been defined.

Such, then, is our formal conclusion, but human affairs – and often with good reason – do not necessarily follow the line of logical development appropriate to the case. Uniformly throughout the civilized world, for two thousand years and more, independent (primary) units of volume have been defined and legalized for the purposes of trade. Generally, standards were set up in the form of cylindrical vessels – vessels which could be filled level to the brim with grain or wine. We should not here be concerned with these units, except that one of them has been widely used for the purposes of scientific measurement – particularly by the chemists, since the days of the French revolution. In 1799, having defined the metre (p. 27), the National Assembly sought to define the unit of mass (see p. 132) as the mass of one cubic decimetre (1000 cm^3) of pure water at its temperature of maximum density (4°C). Because the volume of a vessel of arbitrary shape can most conveniently be deduced by filling it with water and weighing the contents, it was not unreasonable to give a special name to the volume occupied by unit mass of pure water (at 4°C). The name given was the *litre*. According to the definition of 1799 the litre was 10^{-3} m^3, precisely. From that time on, analytical chemists have tended more and more to make up their standard solutions in terms of the litre as unit.

When the definition of the kilogramme, as unit of mass, was changed in 1889 (destroying its relation to the unit of volume), the status of the litre became ambiguous. It could not (by definition) at the same time be 10^{-3} m^3, precisely, and also the volume occupied by one kilogramme of pure water at 4°C. The position was clarified by decree in 1901: the litre was to be the volume occupied by one kilogramme of pure water at 4°C and standard atmospheric pressure – and it turned out that this volume was in fact 1000·028 cm^3 (to one part in 10^6). In 1964 the twelfth General Conference of Weights and Measures abolished the litre as an independent unit. The name could be retained, so it was agreed, by those accustomed to its use, as a special name for a volume of one cubic decimetre, but its definition in terms of the unit of mass was abandoned.

In the last section we dealt at some length with the problem of angular measurement, restricting our discussion to the plane angle of two-dimensional geometry. Here, very briefly, we refer to the measurement of 'solid angle'. 'Estimation' or 'specification' would indeed be a more appropriate word than 'measurement' in this connexion, for there is no instrument available by which the measure of a solid angle can be read off on a scale.

The simplest realization of a solid angle is at the common meeting point of three planes, as, for example, occurs at each corner of a tetrahedron. The measure of the solid angle enclosed by three planes is the measure of the 'openness' of the corner in which they meet. Imagine a sphere, of radius r, to be described about the point of intersection of the planes as centre; then if an area A of the surface of the sphere is enclosed between them (A is the area of a spherical triangle in this case), ω, the measure of the solid angle concerned, is given by the defining equation $\omega = A/r^2$. Here we have described a special case. An alternative definition is at once more general. We say that if there be a closed curve of arbitrary shape lying wholly in the surface of a sphere of radius r, and if this curve encloses an area A of that surface, then the solid angle subtended by this area at the centre of the sphere is given by A/r^2. In that case, if a radius is drawn from the centre of the sphere to any point on the curve, then, when this point traverses the curve, the radius generates the surface of a cone having its apex at the centre of the sphere, and the solid angle in question is the solid angle at the apex of the cone.

Here, we make no further comment, except to say that the magnitude of a solid angle is given by a pure number, the ratio of two areas, and that the 'natural' unit, represented by $A = r^2$ in the defining equation, is the *steradian*. A complete sphere subtends a solid angle of 4π steradians at its centre.

Further reading

F. Bitter, *Mathematical Aspects of Physics*, Doubleday, 1963.

M. Jammer, *Concepts of Space*, Harvard University Press, 1954.

A. N. Whitehead, *An Introduction to Mathematics*, Oxford University Press, 1911.

Units and Standards of Measurement employed at the National Physical Laboratory, *I*, H.M.S.O. (revised periodically).

33 Area, Volume and Solid Angle

Chapter 3
Time

3.1 Introductory

We have said that 'it is the physicist's business to unravel the complexities of phenomena [and so] to lay bare the underlying regularities' (p. 14) – and, in another place (p. 18), 'physics is concerned with the discovery of what is permanent in the changing appearance of the world'. This antithesis between permanence and change is meaningless except in the context of the notion of time.

'Time is the abstract of all relations of sequence,' wrote Leibniz, so complementing his other statement, which we have already quoted (p. 18), 'space is the abstract of all relations of co-existence.' These are neat forms of words, and they have the ring of truth in them, but they do not advance the understanding of the physicist in any positive way. In respect of the problem of time, the physicist must work out his own salvation – his own empirical solution – undeterred by the difficulties. The difficulties, indeed, are considerable: the notion of time is at once a private and a public notion – subjective and objective. The physicist, though he lives with and through his own private awareness of time, past and present and in the imagined future, as professional physicist must strive always for complete objectivity: he must strive to render measurable public time passing in the present.

In relation to the measurement of length, it is axiomatic for practical geometry that a graduated scale may be moved from place to place without change. The same reference length (scale interval) may be used for any number of successive operations of measurement. In relation to the measurement of time, on the other hand, a reference interval may be used only once. It is not possible to construct an equally divided time scale by moving a chosen time interval forward so that its beginning coincides with the previous position (in time) of its end, repeating the process indefinitely – as, in practice, it is possible to construct an equally divided length scale. One has to look for repetitive processes in nature (or devise them in the laboratory) – processes in which uniform patterns of recurring features may be discerned. Then, once the assumption has been made that the recurrence interval is constant, such a process may be used – if only tentatively – to provide a natural, or arbitrary, time scale, as the case may be.

Without naming it, we have just quoted (at the beginning of the last paragraph) the 'axiom of transferability' in relation to material scales in the practical geometry of Euclid. There is another belief which is wider in its field of reference, for it informs the whole outlook of the physicist. Maxwell has expressed it as follows (*Matter and Motion*, p. 21): 'The difference between one event and another does not depend on the mere difference of the times or the places at which they occur, but only on differences in the nature, configuration, or motion of the bodies concerned.' This belief, also, has the character of an axiom: it may be referred to, loosely, as the 'axiom of orderliness' (see p. 13). Obviously, the degree of its usefulness depends on the extent to which we know, in any situation or class of situations, precisely which are 'the bodies concerned'. This is the old problem of the fiction of the isolated system (p. 13). We have to ignore its difficulties – or rely on our intuition to minimize them – before we can use the axiom of orderliness, as we wish to use it, to specify more closely the type of process for which it can be assumed axiomatically that it is a repetitive process for which the recurrence interval is constant. That is the only reason why we have introduced the axiom here: for the purpose of establishing an equally divided time scale we need either to identify a suitable repetitive process on the basis of axiomatic specification, or to attempt to evaluate a series of arbitrary choices by appeal to their consequences. It is the method of experimental science always 'to appeal to the consequences' in the end, but it is useful to examine the axiomatic approach on its merits, nevertheless.

From the point of view of the axiom of orderliness, there are two prime conditions that a process must satisfy if a satisfactory time scale is to be established in terms of it. The process must involve a system of bodies of enduring identity, and it must be such that a particular 'aspect' of the system, having occurred once, occurs again precisely as before. By the term 'aspect' we signify the totality of configuration and motion, in Maxwell's phraseology. Alternatively, we may express the second condition 'operationally' in the longer statement that, if we observe the system, then as the process develops we must be able to recognize a 'second' occasion on which the relative positions of the bodies constituting the system, and their relative velocities, are precisely what they were on a 'first' occasion. The aspect of the second occasion having 'arisen out of' that of the first, a third precisely similar aspect must arise out of the second, and a fourth out of the third, and so on, with constant periodicity, if the axiom of orderliness is applicable to the case. On the face of it, it would seem reasonable to conclude that a process which fulfils these two conditions is indeed a repetitive process for which the recurrence interval is constant.

We have already sounded a word of caution in relation to the involvement, in this axiomatic approach, of the fiction of isolation. If we are to be properly cautious there is another difficulty that we must notice – and attempt to avoid. We remind ourselves that our real reason for seeking to identify a repetitive process of constant recurrence interval is so that we may use it to establish a soundly based system of time measurement. In strict logic, the setting up of such a system must clearly precede the setting up of a system of velocity measurement (p. 47). This being the case, we must obviously be suspicious of any recipe for the identification of the basic repetitive process which makes the assumption that velocity measurement is already possible. As we have developed the axiomatic approach, starting from Maxwell's (necessarily general) statement of the axiom of orderliness, at least in our operational version of the second condition, we have apparently made just this assumption – in this version our second condition states that we should be able to recognize two separate occasions on which the velocities of all the bodies concerned in the process are the same. Escape from the futility of a tautological situation seems impossible, on the basis of this analysis, unless it be admitted that we can recognize the momentary occurrence of lack of relative motion in a physical system without the necessity of measuring velocities. If this 'last chance' solution of our philosophical difficulty is acceptable, we can restate the second condition operationally as follows: observing the process develop we must be able to discover two occasions on which the aspect of the system is precisely the same; in particular, the relative positions of the bodies constituting the system must be the same, and they must be relatively at rest each with respect to all the others, on each of these occasions.

Let us consider an example, for purposes of illustration. We set up a simple pendulum and set it in motion, swinging back and forth in a vertical plane. Very slowly, as we observe the motion against the frame of reference provided by the rigid support of the pendulum, the amplitude of this repetitive process decreases. Twice in each complete cycle there is, instantaneously, a configuration of no internal motion of the system of pendulum and support. If the amplitude did not decrease, but remained constant indefinitely, then, in terms of the axiom of orderliness alone, we should be entitled to assume that the periodic time of the pendulum was likewise constant. We should then have, in the system of the swinging pendulum of constant amplitude, the basis for an arbitrary time scale, equally divided. To make the assumption of constant recurrence interval in respect of an actual pendulum of slowly decreasing amplitude, is to go beyond what the axiom itself allows. It turns out, in fact (see p. 59), that this wider assumption is also justified within limits (when the angular amplitude is small), but in this case the justification comes from experiment – it cannot derive authority from the axiom, though it may be held that its justification by experiment strengthens belief in the axiom itself (see below).

3.2 Practical systems of time measurement

It is said that, in 1581, the young Galileo, paying less attention than was his due to the preacher, observed a lamp swinging in pendulum motion in the cathedral of Pisa, and, using his own pulse for comparison, found to his amazement that the periodic time of swing was the same throughout, although the amplitude changed. Galileo did not significantly exploit this discovery – his classical experiments on motion under gravity (p. 87) were made using water-clocks of traditional design – nevertheless, reversing his original 'experiment', he constructed a simple pendulum device for doctors to use, over strictly limited periods at a time, in determining their patients' pulse rates. Not until the last years of his life, when he was under house arrest, did he come to realize that an oscillating pendulum could more usefully provide an effective means of regulating the motion of a continuously running time-piece, but by then he had not the facilities necessary to put his ideas to the test.

Christiaan Huygens (1629–95) constructed the first practical pendulum clock in 1657. During the years that followed he developed the mathematical theory of pendulum oscillations on a semi-intuitive basis, and showed for the first time that, for small amplitudes of swing, the periodic time of an ideal simple pendulum (p. 59) is proportional to the square root of the length of the suspension. He also made considerable progress towards an understanding of the small-amplitude pendulum oscillations of rigid bodies (p. 145), and obtained an elegant formal result relating to the oscillations of a simple pendulum through large amplitudes. All this, and much more, was contained in his treatise *Horologium oscillatorium*, published in 1673.

In Huygens's day, mechanical clocks, consisting of trains of gear-wheels driven by a slowly falling 'weight', had already been in use for nearly four hundred years. A double problem had to be solved before a pendulum could be successfully incorporated into such a clock as a regulator. An 'escapement' had to be designed which would ensure that the very small distance fallen by the weight was the same in each oscillation period of the pendulum; it had also to ensure that the appropriate fraction of the kinetic energy acquired by the weight was communicated, during each period, to the pendulum, so that its oscillations might be maintained. Previously, there had been no component of the regulating mechanism which was characterized by a definite 'natural period of oscillation'. There had been an escapement of sorts, and a 'balance', roughly in the form of a dumb-bell which could rotate in a vertical plane about a horizontal axis, but this member did not swing back and forth in approximately free motion: it was driven back against gravity and then allowed to 'fall' through a predetermined arc until the next tooth of the escapement engaged the appropriate pallet on its axle. The whole effort producing the retrograde motion of the balance was communicated from the driving weight through the escapement. Obviously, in such an arrangement, variations in frictional effects produced serious irregularities, and good time-keeping was dependent on the constancy of amplitude of the balance motion.

In this last connexion, more than in any other, the introduction of pendulum regulation effected an enormous improvement in reliability generally: the acceptable error in a good clock was reduced from 400 seconds per day to 10 seconds per day, according to the historians.

The ultimate aim of the clockmaker, from the earliest craftsmen who made clepsydras in Athens or Rome, to Huygens himself, was essentially to produce an instrument by which the naturally recurring sequence of daylight and darkness could be subdivided so as to provide a (temporal) framework for the organization of communal life – or the recording of astronomical events. For primitive man the period from noonday to noonday had been 'given', from the beginning, as the natural unit of time reckoning, and his descendants of later millennia, including the clockmakers, instinctively accepted this unit. The clockmakers' aim (at least from the fourteenth century onwards) was to subdivide this period into 'hours' of equal length – twenty-four such hours – and the hours, again, into sixty minutes each, and the minutes into seconds in the same ratio (see p. 26). We say 'from the fourteenth century onwards', because before that time the civic usage of classical Greece persisted: throughout the year the (variable) interval between sunrise and sunset was subdivided into the twelve (equal) hours of daylight and the interval between sunset and sunrise into the (generally different) twelve hours of darkness. In 'classical' times, then, from this point of view, at least, the clockmakers had to accept an essentially more difficult task than that which was given to their successors of the Renaissance.

The philosophers of the earlier civilizations, fixing their frame of reference in the earth, ascribed the alternation of daylight and darkness to the supposed revolution of the sun around the earth, and the changing aspect of the heavens to the revolution of the starry sphere. This was the orthodox viewpoint, though it was not the only one that was considered. In the fourth century B.C., Hicetas of Syracuse, and Ecphantus, had sought to account for the observed common motion of the 'fixed' stars in terms of the steady rotation of the earth on its axis – leaving the sun to perform its revolution around the earth once in a year rather than once each day, the moon once in 27·3 days rather than once in something less than twenty-five hours, and the five planets the more complicated motions which gave them the character of 'wanderers' in the firmament. Again, in the third century, Aristarchus of Samos, abandoning the traditional viewpoint entirely, had maintained that it is the sun that is at the centre of things, with the earth and the planets in revolution around it. These speculative and unorthodox views became the subject of keen debate for a season, but the weight of opinion was against them (as well it might have been, as long as there was no realization that the majority of stars, as seen from the earth, are more than a million times more distant than the sun). Hipparchus, in particular, defended the orthodox viewpoint, though he modified it in detail. As a result of his own observations, over the years from 161 to 126 B.C., from his observatory on the island of Rhodes, he was in a better position than any of his predecessors had been to assess the rival hypotheses, and he asserted with good reason that there was nothing in what he had observed to require the abandonment of the geocentric cosmology of the ancients. Regarding the matter objectively, we may say that on Hipparchus' authority alone the old argument was clinched, in favour of the fixed-earth hypothesis, for well-nigh seventeen centuries more. Hipparchus' cosmology – together with the trigonometry of Apollonius (c. 225 B.C.) which provided its foundation – was later codified by Claudius Ptolemy of Alexandria. Over the period from A.D. 127 to A.D. 151, Ptolemy compiled the thirteen books of the

Syntaxis (later translated into Arabic as the *Almagest*), thereby providing its definitive exposition for the forty generations that followed.

The basic requirement of an acceptable cosmology had been stated by Plato (p. 11): it must 'preserve the appearances of phenomena'. Yet there was one such 'appearance' that had to be discounted from the outset. No progress could be made unless it were assumed that the daytime sky is studded with stars, as the night sky is – stars which are invisible to the terrestrial observer only 'because of the brilliance of the sun'. We have already presupposed this assumption in stating the orthodox view of the ancient philosophers, and in describing the modification of that view which Hicetas suggested. The assumption dates back to times of which no written record remains. Identification of the five 'wandering stars' belongs to the same period of prehistory – and the recognition that the sun and the moon, also, move steadily against the background of the stars. In times of which we have no surviving account, it was already a matter of common astronomical knowledge that the period of one complete traversal of the zodiac by the sun is also the period of one complete cycle of variation of the sun's noonday altitude, the same period as that of the seasonal changes of vegetation on the earth, the year in which life in general has its longest rhythm. Furthermore, long before the birth of Eudoxus (p. 16), the accumulated precision of the observational data was such that the sun's motion through the zodiac was known to be slightly uneven, more rapid in the winter (of the northern hemisphere) than it is in the summer. As a disciple of Plato, Eudoxus had set himself the task of 'preserving the appearances' of celestial events in terms of uniform rotations, exclusively: to match the knowledge of the time he found it necessary to employ a combination of three such rotations to represent the sun's motion, whereas (almost by definition!) he was able to represent the apparent motion of the fixed stars by a single rotation only. The classical philosophers were at one in considering the sphere as the perfect solid, and uniform rotation (or uniform circular motion) as perfect motion; on this basis, if they had been fully consistent, they would surely have preferred to use the motion of the fixed stars, rather than the imperfect motion of the sun in the heavens, as providing their standard of time. But the stars are invisible by day, and the sun rules the lives of men: time measurement for them was an essentially local requirement – the notion of universal time, or physicists' time, had yet to be formulated.

The world-system of Hipparchus and Ptolemy provided the accepted cosmology for the theologians – and for the natural philosophers – of western Europe for some seventeen centuries. Its overthrow took place in the century and a half between 1530 and 1680. In 1530 Nicolaus Copernicus (p. 17) had all but completed the manuscript of a treatise expounding a heliocentric world view which was finally published as *De revolutionibus orbium coelestium* in 1543. Already, by the time of its publication, this cosmology was familiar in outline to many scholars, for it had been widely discussed through oral transmission. This, perhaps, was its chief merit: to awaken interest in the problem, and to demonstrate conclusively that, under the same restrictions, a heliocentric system is essentially simpler than one which is earth centred. But the fact is that Copernicus was content to be bound by the Ptolemaic restrictions. Accepting uniform circular motion as the basic ingredient, he built up his system, in the classical manner, of cycles and epicycles. The measure of his achievement was that he was able to reduce the eighty independent circular motions of the geocentric system of Ptolemy to a mere thirty-four, when the 'origin' was transferred to the sun.

The Copernican cosmology was championed by Giordano Bruno (1548–1600) and by Galileo. For the former it became a necessary, though almost trivial, component in the grandiose speculations which went to the writing of *De l'infinito universo e mondi* (1584); for the latter it provided the occasion for the brilliant polemic of *Dialogo dei due massimi sistemi del mondo* (1632). In the interim between these two publications, each of which helped to bring down upon its author the serious displeasure of the Church, a lapsed seminarist of Lutheran persuasion, Johann Kepler (1571–1630), untroubled by the necessity of obedience to Rome, was engaged on a new attempt to 'preserve the appearances of phenomena' (p. 39), adopting for that purpose only the heliocentric viewpoint of Copernicus and discarding the rest. Kepler was fortunate in having at his disposal the systematic observations of Tycho Brahe (1546–1601), to whom he had acted as personal assistant for a period immediately before his death. For the first time, for more than a thousand years, observational material more accurate than that of Hipparchus had been available to a mathematician of ability and insight. Kepler made good use of this material. In *De motibus stellae Martis*, published in Prague in 1609, he announced the first results of his analysis. He had shown to his own satisfaction that the motions of the planet Mars and of the sun, as seen from the earth, could be satisfactorily accounted for, in a frame of reference in which the sun was at rest, by ascribing appropriate elliptical orbits to the earth and to Mars, each orbit being such that one focus was centred in the sun, and the motion of each being characterized by the same law of velocity-variation in its orbit. Ten years later, in *Harmonice mundi*, Kepler showed that the apparent motions of Mercury, Venus, Saturn and Jupiter could be similarly described – and he added a third empirical law to the two which he had enunciated in the earlier treatise. We shall be concerned with Kepler's three 'laws of planetary motion' in a later chapter (p. 148). Suffice here to say that in the heliocentric system which he described the orbits of the planets are simply ellipses, one focus of each having common location in the sun.

Kepler's system was complete with the publication of the third law in 1619. As a high-grade approximation to descriptive 'truth' it is accepted today. Yet we have set the year 1680, or thereabouts, as the time of the final overthrow of the system of Ptolemy. There are two reasons for choosing this later date. In the first place, Kepler's work was altogether ignored by Galileo: thirteen years after the publication of *Harmonice mundi*, Galileo was concerned only to argue in favour of the heliocentric system of Copernicus. In the second place, and more cogently, during the half-century which followed its publication, Kepler's system merely had the merit of descriptive simplicity. Although this was indeed a great merit, it was not until Newton provided the dynamical basis for the understanding of the empirical laws, whereby simplicity of description became merged in a fundamental simplicity of interpretation, that the case for the acceptance of the Keplerian cosmology became unanswerable. It appears that Newton reached this interpretative goal first about the year 1677. In 1684 and 1685 he brought it to the stage of formal mathematical demonstration. We shall be concerned with Newton's contribution in Chapters 6 and 7; here we have sufficiently justified our choice of the decade around 1680 as the period of final abandonment by men of science of a descriptive cosmology involving cycles and epicycles, whether we consider the cosmology of Ptolemy or the world-system of Copernicus in this connexion.

For the physicist's purpose of time measurement, only two repetitive processes were revealed by the new cosmology to be earth related: the rotation of the earth on its axis, and its orbital revolution in its elliptical orbit around the sun. Obviously, the former of these is essentially simpler in character than the latter. It was, therefore, an obvious choice for a repetitive process of assumed constant recurrence interval. Following the procedure of evaluation 'by appeal to the consequences' (p. 35), the physicist – or astronomer – had then to examine the question experimentally whether it was consistent to make the double assumption that the earth's rotation and the small-angle oscillations of a pendulum of constant length are, each of them, repetitive processes of this character. The introduction of the 'transit circle' by Tycho Brahe, and its further development by Roemer (see p. 166), about 1690, eventually made this possible. A transit circle consists essentially of a telescope mounted so as to be capable of rotation about a horizontal axis fixed in an east–west direction. In this way the passage of any star from east to west across the meridian may be observed, and the interval of pendulum-clock time between successive transits may be recorded. The accumulated evidence of observations of this character on a multitude of stars over a period of two hundred years appeared to show that, as successive refinements of design were introduced into the clock mechanism, the recorded interval of pendulum-clock time between successive star transits became more nearly constant over all. By 1890, with the best astronomical clocks, the residual error (ascribed to the clock) was of the order of 10^{-2} second per day (see p. 37), or roughly one part in 10^7. For all practical purposes, therefore, the double assumption had been justified, and the physicist could reasonably believe that a system of time measurement based on pendulum oscillations was a well-founded system. Throughout the previous century, and longer, using essentially this system, he had already elicited many regularities of striking simplicity among natural phenomena – natural 'laws' capable of representation by simple mathematical symbolism in which the time concept was involved: that was an earlier source of confidence in the system, though of a different character.

We have referred more than once, already, to the non-uniformity of the apparent motion of the sun through the firmament. Because of this fact, if the

41 Practical Systems of Time Measurement

transit-circle experiment is carried out with the centre of the sun's disk effectively taking the place of a star, the interval of pendulum-clock time between successive transits (or 'southings') is found not to be constant; instead it varies cyclically throughout the year. This interval – the solar day – has a maximum variation from its mean value of about one part in 3000 (this occurs in mid-December, when the solar day is at its longest). Even Huygens's pendulum clock (p. 37) could have detected this variation if it had been kept going, in a constant environment, over a period of a few months. On the other hand, until the end of the nineteenth century, as we have already implied, no variation in the length of the sidereal day – the interval between successive southings of any of the fixed stars – had been set in evidence by the use of pendulum clocks and transit instruments in the way that we have described.

Developments in time-keeping during the twentieth century have modified this situation. In 1924 W. H. Shortt introduced the free-pendulum arrangement for controlling the rate of a weight-driven 'slave' clock. In this arrangement, the 'free' pendulum is the essential time-keeper. It oscillates in an evacuated case maintained at a constant temperature, and every half-minute receives a small mechanical impulse which makes good the residual energy loss due to 'friction'. Connexion between this master pendulum and the slave clock mechanism is through an external electrical circuit. The accuracy is of the order of 4×10^{-3} second per day. The intercomparison of the times recorded by two Shortt clocks gave the first indication that these clocks kept in time with one another more closely than either 'kept in time with the stars'.

More recently even more reliable time-keeping has been achieved by the so-called 'quartz clocks' and 'atomic clocks'. Any regular solid body has certain fundamental vibration-modes in which it is capable of elastic oscillation with definite frequencies. Quartz crystals are also 'piezoelectric', that is they become electrically polarized when subject to mechanical stress – or, alternatively, alter dimensions slightly when placed in an electrostatic field. On this basis the mechanical oscillations of a quartz crystal may be maintained electrically, and such an oscillating crystal may be used as the controlling unit of an electric oscillator. For precision use, the crystal is housed in a small, sealed metal box at constant pressure in a constant temperature enclosure. Over not too long periods, such a system can be made reliable to within 2×10^{-4} second per day.

The first atomic clock was developed at the U.S. Bureau of Standards in 1949. Basically, an 'atomic' clock makes use of a natural frequency characteristic of an isolated atom or molecule. In a gas at low pressure, or in a beam of atoms or molecules emitted from a high-temperature enclosure ('oven') through a suitable arrangement of apertures into an evacuated space, individual atoms or molecules may be regarded as free from mutual interaction. In such conditions of isolation, if an atom or molecule emits energy in the form of electromagnetic radiation, or absorbs energy from the electromagnetic field in its neighbourhood, the radiation involved cannot be of any arbitrary frequency, but is sharply restricted to one of a series of frequencies characteristic of the atom or molecule concerned. We have already seen how this empirical 'fact' has been made use of in the 1960 redefinition of the metre in terms of the wavelength of the orange radiation from a low-pressure gas discharge in krypton-86 (p. 28). For the purposes of an atomic clock, absorption of radiation by the isolated atom or molecule, rather than emission, is the fundamental process involved – and a frequency very much less than that of visible light is obligatory. The essential equipment involves a 'microwave' oscillator whose frequency can be varied over a small range through the characteristic atomic frequency which it is intended to employ. When the oscillator is correctly tuned to this characteristic frequency, absorption of electromagnetic energy in the 'absorption cell' containing the low-pressure gas or the atomic beam is at a maximum, and by the use of appropriate control circuits the oscillator can be made to 'lock' on this frequency. To drive the 'dial' which displays or records the 'time', electrical energy is taken from the system at a frequency which is a convenient sub-harmonic of the oscillator frequency.

The first atomic clock of 1949 employed a characteristic frequency of the ammonia molecule, NH_3. This frequency was known to be in the region of $2 \cdot 3870 \times 10^{10}$ Hz in terms of the mean solar second as the unit of time. It is the frequency characteristic of an 'inversion' of position of the nitrogen atom (ion) in the ammonia molecule. The accuracy of this clock was estimated as some three parts in 10^9 over periods of moderate duration.

In 1955 an atomic clock using a beam of caesium atoms in vacuum was successfully operated for the first time. The characteristic frequency employed, approximately $9 \cdot 192 \times 10^9$ Hz, is a 'hyperfine structure' frequency of the caesium atom in its ground state. It characterizes a reversal of the magnetic moment of the atom brought about by a change in direction of the spin axis of the valency electron. A detailed study of the behaviour of such caesium clocks showed that time intervals are reproducible, by their aid, to three parts in 10^{11} under favourable conditions. In the relatively few years during which these clocks have been used by astronomers, short-term variations of the rate of the earth's rotation, which previously had required the detailed reduction of less accurate observations spanning several decades for their mere identification, were set in evidence unambiguously.

By international agreement, at the twelfth General Conference of Weights and Measures in 1964, the caesium clock was accepted for interim use in the definition of a natural unit of time. It was also agreed that before such a definition was made permanent, further research should be undertaken into the possibility of there being atomic frequencies more suitable as reference frequencies than that of the electron-spin transition in caesium, and physical systems capable of matching such frequencies with greater stability than that of the magnetic resonator in the atomic beam arrangement of the caesium clock. In a sense all this was justifiable caution – in 1964 experimental techniques in the relevant fields of radiation physics were currently developing with great rapidity – but four years previously the unit of length had been given 'final' definition in terms of the wavelength of an atomic radiation, and the time unit of the interim definition of 1964 was already capable of realization in practice with a relative accuracy greater than that with which the length unit could be realized. It might well be thought that the matter could have been clinched there and then. In fact, more was involved than the desirability of surveying other possible reference frequencies, as will now appear.

Basically, the principle of the redefinitions of 1960 and 1964 was to alter the specification of the standard (of length or of time) without significantly altering the size of the practical unit. This aspect of the matter has already been fully discussed in relation to the redefinition of the unit of length in the last chapter (p. 31). In that case the situation was clear cut. Not since 1889 had there been any ambiguity in the operational definition of the length unit (metre): a change of standard – from a platinum-iridium bar to a radiating atom – could therefore be contemplated as a simple issue without complications.

It was otherwise, in 1964, in relation to the unit of time. The operational definition of the second in terms of phenomena associated with the rotation of the earth is essentially difficult, for reasons that we have already partially exposed; in 1964 the most recently identified of these difficulties had only just been taken count of in a new definition. The 'ephemeris second' had been defined (in 1956) – we paraphrase the definition here for sake of easier intelligibility – by the statement that the interval between the vernal equinox of the year 1899 and that of the following year 1900 was 31 556 925·9747 ephemeris seconds. (The vernal equinox is the instant in the yearly cycle when the centre of the sun, moving from south to north against the background of the stars, is precisely 90° from the celestial pole 'as seen' from the centre of the earth.)

The difficulty that was recognized in this definition was the logical difficulty attaching to the traditional (implicit) assumption that the rate of rotation of the earth is indeed invariable (such an assumption effectively excludes this particular motion from the scope of the laws of dynamics). Definitions of the second based on the mean solar day – or the mean sidereal day – inevitably involve this difficulty. In the definition of 1956 the difficulty was resolved by replacing the rotation of the earth on its axis by its revolution in orbit around the sun, as the basic repetitive process – and, so that the difficulty should not recur in the new context, by specifying just one particular circuit of that orbit (that containing 1 January 1900), leaving to the terrestrial clockmaker the real practical problem of maintaining the effective repetitive standard.

Two comments may be made on the 1956 definition of the ephemeris second. In the first place we note that the 'tropical year' 1900 is specified (a tropical year is an interval between successive vernal equinoxes). The reason for this backwards reference is by no means exclusively the sentimental appeal of the 'round number'; more practically it issues from the fact that the complete numerical reduction of the necessary astronomical observations is a lengthy process, so that, at least before the introduction of the computer, precise knowledge of the actual relative motions of the earth and the sun was regularly several years in arrear. In 1956, 1900 was a relatively recent year, but sufficiently distant in time that data in respect of it were unlikely to be further modified by adjustments based on later calculations.

Our second comment relates to the precision of the definition. We note that twelve significant figures are given – and we recall that even today it is impossible to make a laboratory determination of frequency, in an experiment of tolerable duration, to better than about one part in 10^{10} (p. 43). Here we have to admit that the concern of the astronomer – as distinct from the physicist – in this regard is not limited to 'laboratory experiments of tolerable duration'. The astronomer may be interested in evidence regarding the rotation of the earth drawn from records of eclipses of the sun made by the ancients. On the assumption that the period of the earth's rotation has remained constant, he can calculate the time of day corresponding to totality for each eclipse. It is not difficult to show that if the length of the day has in fact been changing regularly, over the past 2000 years, by as little as one part in 10^{12} per day, then, corresponding to that remove in time, the astronomer's calculations would be in error by nearly seven hours. Essentially qualitative evidence may not be insignificant, therefore – evidence as to whether a particular eclipse occurred in the morning or the afternoon – in relating to matters of such high precision.

In 1964 the acceptance of the caesium clock as an interim time standard was made in relation to the definition of the ephemeris second of 1956. To this end the frequency of the hyperfine-structure transition of the caesium atom in its ground state 'unperturbed by external fields' was conventionally designated as 9 192 631 770 Hz (reciprocal ephemeris second). The number of significant figures in this designation will be seen to be realistic in relation to the physicist's use of the time variable in experiments which involve the ultimate in attainable precision: there has been no attempt to follow the astronomer to the twelfth figure of the definition of the ephemeris second itself.

In conclusion, then, we summarize the gist of our argument: the physicist now has at his disposal various clocks, each capable of being constructed in a well-equipped laboratory according to a precise specification, and each making use of a different repetitive process, and his observations, comparing one clock against another, sustain his belief that, to an accuracy of one part in 10^9 or better, the repetitive process in each case can be significantly regarded as of constant recurrence interval. To the same accuracy, the repetitive process which is the rotation of the earth is demonstrably not such a process; furthermore, in principle the deviations from constancy in this case can be understood in terms of the laws of dynamics (see p. 135). It is a matter of no surprise, in the end, that the repetitive processes which appear genuinely invariable in the matter of frequency are processes characteristic of individual atoms or molecules in a rarefied gas: an individual atom, in such circumstances, approximates more closely to the physicist's ideal of an isolated system than any gross body is likely to do.

45 Practical Systems of Time Measurement

Further reading

R. Schlegel, *Time and the Physical World*, University of Michigan Press, 1961.
S. Toulmin and J. Goodfield, *The Fabric of the Heavens*, Harper & Row, 1961.
G. J. Whitrow, *The Natural Philosophy of Time*, Nelson, 1961.
Units and Standards of Measurement employed at the National Physical Laboratory, I, H.M.S.O. (revised periodically).

Chapter 4
Motion

4.1 Introductory

In the last chapter we adumbrated the notion of time. In preface we stated (p. 34), '[The] antithesis between permanence and change [in the physical world] is meaningless except in the context of the notion of time.' Philosophers may prefer to write, 'The physicist's notion of time is without foundation except in terms of the analysis of the changes which he observes in his physical environment and, in particular, of those changes to which he applies the term *motion*.' Aristotle exemplified this point of view when he wrote of time as 'the number of motion'. We have to concede the validity of this assessment: all that we have written in the last chapter concerning the choice of a time standard has had reference to the relative merits of different physical systems in respect of the stability of the recurring pattern of motion (or change) which is characteristic of each. Indeed, throughout our discussions of space and time as concepts in their own right, we have been unable to avoid constant reference to the idea of movement: at least at the macroscopic level, the world does not stand still for our passive contemplation. Even in our discussions of the Euclidean geometry of space, the idea of movement enters. Now, in this chapter, we direct our attention specifically to its description, in the terms in which the classical physicist has sought to describe it since the days of Galileo, eschewing further philosophizing. We merely remark that we do so against the background of concepts of space and time as continua: space a three-dimensional continuum, and time one-dimensional – and each of infinite extent. This being our assumption, motion – change of position in space – is itself essentially continuous. Nothing in our experience of the behaviour of macroscopic bodies belies this conclusion. For sake of simplicity we shall discuss first the motion of a particle (see p. 19); afterwards we shall enlarge our categories of description in order to include the motion of extended bodies. In neither case, at this stage, shall we be concerned with the 'causes' of motion: our subject is 'kinematics', simply – it merges into 'kinetics' only with the introduction of the concept of force.

4.2 Linear motion: speed, velocity and acceleration

4.2.1 *General considerations*

Granted the assumption of the continuous nature of space and time, the most general motion of a particle in space is necessarily *linear* motion; the path traversed by the particle is necessarily continuous. Strictly, then, in relation to particle motion, the title of this section is tautological. Nevertheless, we retain it: at some stage we shall find it desirable to distinguish between rectilinear and curvilinear motion; conveniently our title subsumes both.

In general, the (curvilinear) path of a moving particle cannot be completely determined by direct observation. High-speed cine-photography, for example, merely gives a series of positions occupied by the particle at successive instants of time. Moreover, these instants are not themselves precisely determined (p. 18). It is no more than matching formalism to reality, therefore, to consider any path, to a first approximation, as made up of a series of finite displacements occupying finite time intervals – and the actual path of the particle as the locus ultimately obtained when, through higher-order approximation, the length of each time interval becomes (in thought) vanishingly small. In such a process of proceeding to the limit it is convenient, though obviously unnecessary, to assume that the individual elementary displacements are rectilinear.

The notion of a rectilinear displacement has already been formulated in section 2.3. In that section we introduced the vectorial representation of the configuration of a system of particles at a particular instant. In respect of an arbitrary origin in a three-dimensional diagram, such a configuration was represented by a series of position vectors drawn from that origin. We described these vectors as specifying the 'displacements' of the various particles from the origin; we showed that the vectors obtained by joining the ends of any two such position vectors similarly specified the displacement which would transfer the 'first' particle to the position of the 'second' so concerned; finally (p. 24) we argued to the conclusion that the configuration in question was completely specified by 'the displacement vectors relating every such point in the diagram to every other point' – independently, that is, of the origin chosen.

We may take over these general results, having reference to the positions, at a given instant, of a number of particles constituting a 'system' of interest, and reinterpret them with reference to the positions, at successive instants, of a single particle. Obviously, the primary position vectors drawn from the arbitrary origin must now be numbered in (temporal) sequence – and our interest in the displacement vectors is confined to those joining the ends of consecutively numbered position vectors. Obviously, also, for even a 'first approximation' to the actual path to emerge, the 'density' of position vectors in the diagram must be considerable. It would appear as an almost trivial result in the end – echoing the last clause of the last sentence of the previous paragraph – that this first approximation to the path emerges with equal clarity in the diagram irrespective of the choice of origin, and that, in the limit when the 'density' of position vectors becomes infinite, the only significant feature that remains is the delineation of the actual path in the 'representational space' of the diagram.

In the last two paragraphs we have been concentrating attention on the description of the path, merely requiring that the position vectors should be 'numbered in (temporal) sequence'. Let us now require that these successive position vectors should refer to instants equally spaced in time. Let Δt represent the common (small) duration separating one such instant from the next in sequence. Suppose that r_n and r_{n+1} are the position vectors of the moving particle at the nth and $(n + 1)$th instants in this sequence, respectively. Let Δs_n be the displacement vector corresponding to these two terminal instants. Then (see equation 2.5)

$$\Delta s_n = r_{n+1} - r_n, \qquad\qquad 4.1$$

and we define the (vector) quantity $\Delta s_n / \Delta t$ ($\equiv v_n$) as the mean velocity of the particle during the time interval in question. In general, in the limit when $\Delta t \to 0$, we write $\dot{s} \equiv v$, the vector quantity v being defined as the *instantaneous velocity* of the particle at the relevant instant. Clearly, from equation 4.1

$$v = \dot{r}, \qquad\qquad 4.2$$

and, from what we have already said of the essential independence of the vectorial representation of the choice of origin, we conclude that for a given motion the same value of the instantaneous velocity v is given by equation 4.2 from whatever origin the position vectors r are drawn in the representational space that we are using.

The concept of instantaneous velocity, fundamental to Newtonian kinematics, was totally unknown to the philosophers of classical Greece: its ultimate formulation – not rigorously completed until Newton himself developed the 'method of fluxions' – may be seen as foreshadowed in the speculative discussions of a group of scholars at Merton College, Oxford, during the second quarter of the fourteenth century. William Heytesbury (d. 1372) is possibly the best known of this remarkable group of men.

The next step in our formalism is to develop a concept in terms of which we can give a quantitative description of the variation of velocity during particle motion. For this purpose we repeat the process that we have just concluded, drawing a time-ordered sequence of velocity vectors v_n from an arbitrary origin in representational space. In this way we obtain a first-order approximation to a diagrammatic representation showing how the instantaneous velocity of the moving particle changes ('smoothly') from point to point in its actual path. As before, we must enter the caveat that we shall not indeed obtain even a first approximation to an adequate representation of these changes unless the 'density' of velocity vectors in the diagram is considerable. In the limit (in thought) when the 'density' of velocity vectors becomes infinite, the continuous curve obtained by joining the ends of successive vectors in the temporal sequence, taken together with the corresponding curve in the diagram of positions, provides the full representation of the velocity changes concerned. To make this representation complete we have, of course, effectively to mark a time scale along each curve: only in this way may corresponding points along the two curves be identified.

Having repeated the process of constructing a diagram in representational space, let us now repeat, with appropriate modification, the various steps of mathematical formalism which led to equations 4.1 and 4.2 of our previous treatment. The elementary vector obtained by joining the ends of the velocity vectors v_n and v_{n+1} may be referred to, with self-evident justification, as a 'velocity increment' vector: let us denote its value by Δu_n. Then, in conformity with equation 4.1, we have

$$\Delta u_n = v_{n+1} - v_n. \qquad\qquad 4.3$$

If Δt again represents the common (small) duration separating the nth and $(n+1)$th instants in our representational sequence for all values of n, we define the (vector) quantity $\Delta u_n / \Delta t$ ($\equiv a_n$) as the mean acceleration of the particle in the relevant interval. Similarly, in general, when $\Delta t \rightarrow 0$, we write the defining equation $\dot{u} \equiv a$, designating the vector quantity a as the 'instantaneous acceleration' of the particle at the instant in question. Clearly, then, in this case (compare equation 4.2),

$$a = \dot{v}. \qquad\qquad 4.4$$

We have spoken of the two curves that we have so far constructed in relation to the continuous linear motion of a particle in the three-dimensional space of experience: the curve which we constructed in the representational space of position vectors, and the corresponding curve in the representational space of velocity vectors. We may refer to the first curve as the *virtual path* of the particle; the second is commonly referred to as the *hodograph* of the path. The notion of the hodograph was first introduced into mathematical physics by William Rowan Hamilton (see p. 21) in 1846. Its widespread adoption and use can probably be traced to the prominence accorded to it by Peter Guthrie Tait (1831–1901) and William John Steele in *Dynamics of a Particle* published in 1856. As is clear from a comparison of equations 4.2 and 4.4, the essential property of the hodograph may be expressed in the statement, 'the velocity of the representative point in the hodograph is the acceleration of the particle in its path.' Obviously, in this connexion we are regarding the problem of the 'scale' of the hodograph construction as trivial, and fully accounted for.

We have developed the concept of acceleration to take account of the fact that in the real world we have to do with systems in which individual particle velocities vary with time. It is equally a fact that, in many situations of interest, particle accelerations vary with time, also. Obviously, we could continue the process on which we have been engaged in this section, constructing a curve which is the locus of the ends of successive acceleration vectors drawn from an arbitrary origin in representational space, and defining a new vector quantity which is the time derivative of the vector acceleration. Indeed, there is no logical end to the sequence of successive constructions of this type. That we do not continue the sequence farther than we have already taken it is dictated, then, neither by irresolution nor by logic, but by experience: the basic laws of kinetics, so we discover – or so Newton discovered for us – are expressible in terms of the accelerations of particles, exclusively; no other time derivative of velocity (or position) is fundamentally significant in this connexion.

We have just stated that when we come to introduce the concept of force into our discussions, and consider the 'causes' of motion, as we shall do in Chapter 6, we shall find the kinematical concepts of velocity and acceleration as all-sufficient. There is a sense in which, in this statement, and in the way in which we have introduced the concept of acceleration in this section, we have done less than justice to a problem of historical interest in kinematics itself. Having already taken over the formalism of the three-dimensional representation of the configuration of a system of many particles at an instant of time in order to develop the concept of velocity for a single particle tracing out a configuration of successive positions in the course of its continuous linear motion in space, we were able to present it as a 'natural' extension of the method to develop the concept of acceleration as we in fact developed it. In so doing we ended up with a definition which, in respect of rectilinear motion, identifies the 'simplest' situation as that in which the 'accelerated' particle acquires equal increments of speed in equal intervals of time. (In rectilinear motion we are concerned only with the magnitude and 'sign' of the velocity, which thereby loses its full vectorial character: when we are concerned with the magnitude only, irrespective of direction in space, or even of 'sense' along a straight line, we use the term *speed* to denote the 'intensity of motion'.) Now the followers of Aristotle (such as Strato of Lampsacus in the third century B.C.), when they thought of the simplest type of non-uniform rectilinear motion, thought instead of motion in which the times taken for successive equal distances decrease in regular progression, and even after the emergence of rudimentary ideas concerning instantaneous velocity (p. 49) there were those who, following the Greek tradition, regarded the simplest situation as that involving equal increments of speed in equal distances traversed. It remained for Galileo, making the bold assumption that free fall under gravity is non-uniform rectilinear motion of the ultimate in simplicity, to undertake the crucial experiment to decide what its kinematical character might be. By 1591 he had convinced himself (see p. 87) that in such motion equal increments of speed in fact accrue in equal intervals of elapsed time, not in equal distances traversed. We have said that we accept our formal definition of acceleration, without attempting to extend it, because it proved to be the basic kinematical concept upon which the theory of Newtonian dynamics was erected: here we confess that through the argument from simplicity (in this particular case, that there is an important class of natural situations in which particle accelerations, as we have defined them, are constant) there was evidence, before Newton, for the real-world significance of acceleration as the time derivative of velocity. We shall leave it at that.

4.2.2 *Particular cases*

(a) *Rectilinear motion.* In this, and in the following subsections, we shall be considering certain simple cases of linear (particle) motion in their more formal aspects. Because the reader is likely to be familiar with many of the results involved, our treatment need generally be no more than brief: we are largely concerned at this stage merely to lay the foundations for the future discussion of dynamical problems.

We are agreed that the simplest type of rectilinear motion is that in which the acceleration, as defined by equation **4.4**, is constant. In that case, dropping the vectorial notation since we are dealing with a one-dimensional situation, we have, on a single integration,

$$v = v_0 + at. \tag{4.5}$$

In this equation v_0 represents the initial velocity ($t = 0$), and a further integration on the basis of the definition of equations **4.1** and **4.2** gives

$$s = v_0 t + \tfrac{1}{2}at^2, \tag{4.6}$$

if we postulate that $s = 0$ when $t = 0$. Equations **4.5** and **4.6** together yield

$$v^2 = v_0^2 + 2as. \tag{4.7}$$

Equations **4.5** to **4.7** are sometimes referred to as the simple kinematical equations. They are valid, of course, only for uniformly accelerated one-dimensional motion (see p. 81). Galileo's experiments, to which we referred in the last subsection, essentially consisted in the verification of equation **4.6** for motion under gravity with $v_0 = 0$.

It is interesting at this point to consider the alternative views as to the nature of the simplest type of accelerated rectilinear motion which had been current before the time of Galileo (see p. 51) and which his experiments finally discounted. We take first the view that the simplest type of such motion is that in which the speed increases uniformly with the distance traversed. In that case equation **4.5** is replaced by the defining equation

$$v = v_0 + \frac{s}{\tau}, \tag{4.8}$$

the 'linear acceleration', a, of the previous definition being replaced by the 'characteristic time', τ. We have written the equation in this form to emphasize the fact that each term on the right-hand side of the equation must represent a component of velocity. The second term, in particular, because, by assumption, it involves the measure of the (variable) distance in the numerator, must involve the measure of a (constant) time interval in the denominator. We wish to integrate, as before, imposing the same initial conditions, $s = 0$, $t = 0$. First, we rewrite equation **4.8** in the form

$$\frac{ds}{dt} - \frac{s}{\tau} = v_0. \tag{4.9}$$

Then, we note that

$$\frac{d}{dt}\left[s \exp\left(-\frac{t}{\tau}\right)\right] = \left(\frac{ds}{dt} - \frac{s}{\tau}\right) \exp\left(-\frac{t}{\tau}\right),$$

so that, from equation **4.9**,

$$v_0 \exp\left(-\frac{t}{\tau}\right) = \frac{d}{dt}\left[s \exp\left(-\frac{t}{\tau}\right)\right].$$

We may now integrate, having separated the variables. We obtain

$$s \exp\left(-\frac{t}{\tau}\right) = c - v_0\, \tau \exp\left(-\frac{t}{\tau}\right),$$

c being a constant of integration. Inserting our initial conditions, we have, finally,

$$s = v_0\, \tau \left[\exp\left(\frac{t}{\tau}\right) - 1\right]. \tag{4.10}$$

We note, first of all, an important difference between the forms of equations **4.6** and **4.10**. Equation **4.6** yields finite values of s for $v_0 = 0$ for all non-zero values of a. Equation **4.10**, on the other hand, yields $s = 0$ for $v_0 = 0$, whatever the non-zero value of τ. Obviously, in strict logic, motion from an initial state of rest is impossible in this hypothetical case. When $v_0 \neq 0$, we have, from equation **4.10**,

$$s = v_0\, t + \frac{v_0}{\tau}\frac{t^2}{2!} + \frac{v_0}{\tau^2}\frac{t^3}{3!} + \dots,$$

thus, for a given initial velocity v_0, the distance traversed in a given time, according to this kinematical 'law', is greater than the corresponding distance according to the 'law' of constant linear acceleration (equation **4.6**) – if for purposes of comparison we regard v_0/τ as representing the measure of the linear acceleration in this case.

Having discussed the hypothetical case in which the speed increases uniformly with the distance traversed in some detail, we can dispose of the earlier hypothesis of Strato and the Peripatetics more briefly. Strato's hypothesis of simplicity (p. 51) was never expressed precisely. Its most primitive variant may be represented by the equation

$$\frac{dt}{ds} = \frac{\tau}{\sigma + s}, \tag{4.11}$$

(the right-hand member of this equation decreases monotonically as s increases and is finite for all finite values of s, including $s = 0$). Now, by inversion of each of its members, equation **4.11** may be rewritten in the form

$$v = \frac{\sigma}{\tau} + \frac{s}{\tau}.$$

As so written, the equation is formally identical with equation 4.8: we conclude, therefore, that the most primitive variant of Strato's hypothesis is indistinguishable from the hypothesis that the simplest type of accelerated motion is that in which the speed increases uniformly with increasing distance, which we have already discussed. We make no further reference, then, to Strato in this connexion.

The next case of accelerated straight-line motion which we consider formally in this subsection is that in which the acceleration of the moving particle is proportional to the velocity and oppositely directed. In this case we write

$$\dot{v} = -\frac{v}{\tau}, \qquad\qquad 4.12$$

(again, a characteristic time, τ, is involved, if the measure of the time derivative of velocity, $\dot{v}$, is to be given in terms of the measure of the instantaneous velocity, v). After a first integration we have

$$\ln v = c - \frac{t}{\tau},$$

or, if v_0 is the initial velocity, $c = \ln v_0$, and

$$v = v_0 \exp\left(-\frac{t}{\tau}\right). \qquad\qquad 4.13$$

We note that, according to equation 4.13, the characteristic time, τ, is that time during which, at any stage of the motion, the velocity decreases by a factor $1/e$ ($1/2 \cdot 718 \ldots$) – and that formally, at least, the velocity remains finite through infinite time (though for all practical purposes it may be considered to be zero for $t \gg \tau$). A second integration (now in relation to equation 4.13) leads to the result

$$s = v_0\, \tau \left[1 - \exp\left(-\frac{t}{\tau}\right)\right], \qquad\qquad 4.14$$

if $s = 0$ when $t = 0$. According to equation 4.14, the whole motion is confined within the range $0 \leqslant s \leqslant v_0\, \tau$; also, by the time that $t \gg \tau$, the particle has effectively come to rest at the point $s = v_0\, \tau$.

Finally, let us consider the straight-line motion in which there are two components of acceleration, one constant, and the other proportional to the velocity and oppositely directed, which component we have just considered in isolation. We now have

$$\dot{v} = a - \frac{v}{\tau}. \qquad\qquad 4.15$$

In form, equation 4.15 is precisely similar to equation 4.9, except that there is a change of sign in the term involving τ. On the basis of equation 4.10, therefore, if we impose the initial condition $v = 0$, $t = 0$, we have, as a result of the first integration,

$$v = a\tau\left[1 - \exp\left(-\frac{t}{\tau}\right)\right],$$ 4.16

and, after a second integration, with $s = 0$, $t = 0$,

$$s = a\tau t + a\tau^2\left[\exp\left(-\frac{t}{\tau}\right) - 1\right].$$ 4.17

According to equation **4.16**, after a time long compared with τ, the motion in this case is effectively motion with constant velocity $a\tau$ (the 'terminal' velocity) – and, according to equation **4.17**, when that stage has been reached, the distance traversed is less by an amount $a\tau^2$ than it would have been if the particle had moved uniformly throughout with its terminal velocity.

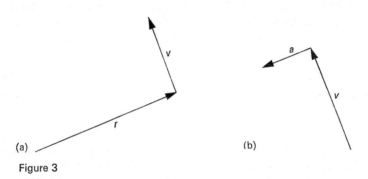

(a)

(b)

Figure 3

(b) *Uniform circular motion.* Suppose that a particle is moving with constant speed v in a circular path of radius r. This is motion in two dimensions, and the hodograph of the path (p. 50) is likewise a two-dimensional curve. Let us suppose that, as we look at it, the path is described in an anticlockwise sense, and let us use the plane of the paper as the common representative space for the two-dimensional vector diagrams of position and velocity (hodograph). Figure 3(a) is an element of the diagram of position, indicating the relative directions of the outwards-drawn radius **r** of the path and the instantaneous velocity **v** at an arbitrary instant. Figure 3(b) is the corresponding element of the diagram of velocity. Here **v** is the (path) velocity vector, and because the hodograph is self-evidently a circle, also described in an anticlockwise sense in the diagram, the arrow labelled **a** may be taken to represent the velocity in the hodograph. Now the (vector) velocity in the hodograph is the (vector) acceleration in the path (p. 50). (That is why we have labelled the hodograph velocity vector **a**.) Our first conclusion from Figure 3 is that the direction of the acceleration in the path is along the inwards-drawn radius – that is, the direction of **a** is the direction of $-\mathbf{r}$. Suppose, now, that the *periodic time* of the motion is τ – the time for one complete circuit of the circular path, or the hodograph. From the geometry of the two curves, we have

$$\tau = \frac{2\pi r}{v} = \frac{2\pi v}{a}.$$

Thus $\quad a = \dfrac{v^2}{r}.$ 4.18

Our final conclusion, in words, is that, when a particle is moving with constant speed v in a circle of radius r, the magnitude of the instantaneous acceleration of the particle is constant and given by v^2/r. As we have already shown, this acceleration is at every instant directed towards the centre of the circular path.

When a particle moves with uniform speed in a circle, the radius vector sweeps out equal angles in equal intervals of time. The length of arc traversed in time t, with linear speed v, being vt, the angle swept out is vt/r (p. 26), and v/r is the *angular speed of rotation* of the radius vector about the centre of the circle. If we represent the angular speed by ω in this case, equation 4.18 may be written in the equivalent forms

$$a = \omega^2 r, \qquad\qquad\qquad 4.19$$
and $\quad a = \omega v.$ 4.20

As they are written – and as they have been developed in the course of our argument – equations 4.18, 4.19 and 4.20 provide no information concerning the vector characteristics of the quantities involved. To supply that information we appended a statement in words specifying the direction of the acceleration the magnitude of which is given by equation 4.18. There is an important sense, however, in which equation 4.20 can be interpreted vectorially, and that without specific reference to the geometry of the circular path to which it refers. Hitherto, we have been considering the problem as essentially a problem in two dimensions: in order to formulate the vectorial interpretation of equation 4.20, we have to see the two-dimensional motion as taking place in three-dimensional space.

In relation to a three-dimensional set of right-handed rectangular axes O X, O Y, O Z, the new interpretation runs as follows: if $\overrightarrow{OX}$ is the direction of the instantaneous velocity of a particle, and if v, the magnitude of this velocity, remains constant while its direction changes with constant angular speed ω in the positive sense about $\overrightarrow{OZ}$ as axis, then the instantaneous acceleration of the particle is in the direction $\overrightarrow{OY}$ and of magnitude given by ωv. It will be observed that we have introduced the third spatial dimension (represented by the axis O Z) in order to specify the 'direction' of the angular motion of which the speed is denoted by ω. We did not write in respect of this motion 'in the direction from X to Y in the plane XOY', but rather, in terms of the right-

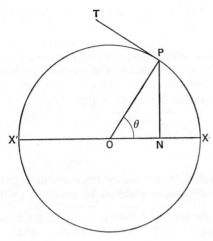

Figure 4

handed screw convention (p. 24), 'in the positive sense about $\overrightarrow{OZ}$ as axis'. Now, we are in no doubt that linear velocity and linear acceleration are vector quantities. In this situation we have been able significantly to associate a direction in space $(\overrightarrow{OZ})$ with an angular motion in a plane (XOY). Arguing from the particular to the general, we opine that *angular velocity* (and *angular acceleration*) are in general vectorially definable – as, indeed, henceforth we shall assume them to be. The 'direction in space' of a rotational motion will be taken to be that direction, at right angles to the plane of the instantaneous rotation, which is positive according to the right-handed screw convention. Obviously, on the basis of this definition, the vectorial characteristics of angular velocity and angular acceleration are not precisely the same as those of linear velocity and linear acceleration. To signify this difference we use the terms *axial vector* and *polar vector* in the two cases.

Equation **4.20**, then, may be interpreted vectorially; indeed, it may be written in vector notation. We shall return to this aspect of the matter when we have accumulated other examples of a similar kind (p. 127). Meanwhile we note that we have written the equation in the form $a = \omega v$, rather than as $a = v\omega$. Apart from the vectorial interpretation, these two forms are interchangeable: their difference consists in the fact that in the former the symbols occur in the cyclic order of the three axes, OY, OZ, OX, with which the respective magnitudes are associated – in the latter this is not the case.

(c) *Simple harmonic motion.* We define simple harmonic motion in a straight line as the projection on a diameter of uniform motion in a circle. Let P (Figure 4) represent the instantaneous position of a particle moving in a circle of centre O and radius A with uniform angular velocity ω. Let X'OX be a diameter of the circle and let N be the foot of the perpendicular drawn from P to this diameter. Then N represents the instantaneous position of a particle which executes simple harmonic motion in X'OX. Suppose that, when $t = 0$, P coincides instantaneously with X, and that the figure is drawn for arbitrary time t. Then, if the motion of P is anticlockwise in the figure, $\angle$XOP $= \omega t$, and x, the instantaneous displacement of N from O, is given by

$$x = A \cos \omega t. \qquad \textbf{4.21}$$
$$\text{Obviously} \quad \dot{x} = -A\omega \sin \omega t \qquad \textbf{4.22}$$
$$\text{and} \quad \ddot{x} = -A\omega^2 \cos \omega t. \qquad \textbf{4.23}$$

Equations **4.22** and **4.23** are most directly obtained by successive differentiation from equation **4.21**; they may also be obtained by assigning to the particle at N the components, in the positive direction $\overrightarrow{X'OX}$, of the actual velocity and acceleration of the particle at P. The instantaneous velocity of this particle is of magnitude $A\omega$ in the direction $\overrightarrow{PT}$ (PT being the tangent to the circle at P), and its instantaneous acceleration is of magnitude $A\omega^2$ (see equation **4.19**) in the direction $\overrightarrow{PO}$. On the basis of these statements, equations **4.22** and **4.23** follow immediately.

Equations **4.21** to **4.23** describe the kinematical characteristics of an oscillatory motion of *amplitude A* about the origin O. According to our assumptions, the periodic time τ of this motion is constant. It is given by the time during which the particle at P makes one complete circuit of the *auxiliary circle*:

$$\tau = \frac{2\pi}{\omega}.$$

The *frequency* of the motion is the number of complete periods per unit time. We may write $f = \omega/2\pi$. Obviously, equations **4.21** to **4.23** may be rewritten by substituting for ω either $2\pi/\tau$ or $2\pi f$.

Although it is convenient to introduce the idea of simple harmonic motion, as we have done, by geometrical specification in relation to a particular case (straight-line S.H.M.), there is a much wider class than this, in the real world, of 'one-dimensional' motions that are simple harmonic. Indeed, in this connexion, the term 'motion' need not be taken literally at all. Literally, motion implies the time change of an 'extensive' variable – position or angle – but the physical quantity which varies simple-harmonically with time may be an 'intensive' quantity such as temperature or the charge on a capacitor. We therefore introduce an alternative, more general, definition, equally applicable to these cases as to the case of rectilinear particle motion that we have already considered. In this development, for sake of conformity, we shall continue to use x, the traditional symbol for a linear displacement, for the time-varying quantity in our equations, but we shall do so without any implication that the interpretation of those equations is limited to that particular application.

Let us note, then, that taking equations **4.21** and **4.23** together we have the result

$$\ddot{x} = -\omega^2 x. \qquad \textbf{4.24}$$

Equation **4.24** provides the basis for our new definition. We say that any physical quantity varies simple-harmonically with time when the second time derivative of the quantity is negative when the quantity itself has a positive value, and is positive when the quantity is negative, provided that throughout the motion the measures of the second time derivative and the quantity (irrespective of sign) bear a constant ratio one to the other. More simply (and possibly more familiarly), in respect of the special case of the rectilinear motion that we have already considered, the new definition reads, 'Rectilinear motion about an origin is simple harmonic when the instantaneous acceleration of the moving particle is all the time directed to the origin and proportional to the instantaneous displacement.'

Equation **4.24** is a second-order differential equation – and equation **4.21** represents a particular solution. The general solution of such an equation must involve two arbitrary constants, and each of these may take any value, as appropriate to the circumstances of the particular case. The general solution of equation **4.24** is

$$x = A \cos(\omega t + \delta). \qquad \textbf{4.25}$$

The time-varying quantity $\omega t + \delta$ is referred to as the instantaneous value of the *phase angle* of the motion, or more simply of the *phase*; the constant quantity δ is the phase angle at the zero of time. δ is sometimes called the *epoch* of the motion. Whatever the values of A and δ, provided ω is the same, the periodic time is the same ($\tau = 2\pi/\omega$) for all simple harmonic motions represented by equation **4.25**. Physically, we may say that if there is any situation to which equation **4.24** applies, with ω constant – and, indeed, there are innumerable such situations, as we shall presently discover – then, in any individual such case, the natural oscillations of the system are of constant periodic time irrespective of amplitude, they are characterized by a 'natural frequency' which is peculiar to the system. We recall that the young Galileo observed this property of *isochronism* in relation to 'simple-pendulum motion' in the cathedral at Pisa in 1581 (p. 36).

In this chapter we are concerned generally with the description of motion, and in particular with types of motion which are exemplified in the real world. For such types of motion it is frequently the case that alternative descriptions are possible and that, depending on the problem in hand, one of these descriptions is mathematically more convenient, or physically more illuminating, than the other. Starting from this viewpoint, we consider here certain examples of equivalent descriptions (which are more commonly to be found, in textbooks on kinematics, under the heading 'composition and resolution of simple harmonic motions').

We start by considering the case of uniform motion in a circle of radius A with angular velocity ω. With reference to the centre of the circle as origin and any diameter (see Figure 4), in this case the polar coordinates of the moving particle are (A, θ). With a suitable choice of time-zero, therefore, the motion of the particle is completely represented by the equation

$$\theta = \omega t, \qquad\qquad\qquad 4.26$$

A being constant. Suppose, now, we seek to describe this motion in terms of the Cartesian coordinates of the particle. If we take the diameter previously used as reference axis of direction as the axis of x, and if the motion is anti-clockwise, the Cartesian coordinates are $(A \cos \theta, A \sin \theta)$ and we have

$$\begin{aligned} x &= A \cos \omega t, \\ y &= A \sin \omega t = A \cos(\omega t - \tfrac{1}{2}\pi). \end{aligned} \qquad\qquad 4.27$$

In this alternative description there are two independent statements – and we recognize that the 'actual' motion has been *resolved* into two rectilinear simple harmonic motions taking place in directions at right angles to one another, each centred in the origin and of amplitude A, each of the same period $2\pi/\omega$, but one (the x-component S.H.M.) 'leading' the other by a phase difference of $\tfrac{1}{2}\pi$. Moreover, since the choice of rectangular axes for our alternative description was entirely arbitrary, we conclude that the originally described circular motion might equally well have been similarly resolved into simple harmonic motions along any two diameters of the circle which are mutually at right angles. In every significant particular, then, uniform motion in a circle is equivalent to two coexisting ('superposed'), mutually perpendicular, rectilinear simple harmonic motions of the same amplitude and period, having the same origin and a constant phase-angle difference of $\tfrac{1}{2}\pi$. Which of these two simple harmonic motions leads the other in phase is obviously determined by (or determines) the sense in which the circular path is described.

In considering uniform motion in a circle we have been dealing with a special case of motion in an ellipse. A circle is a *degenerate* ellipse (in the sense that the two foci are no longer distinguishable, being coincident in the centre of the circle, with the result that every diameter is of the same length). Associated with any circle of diameter $2A$ there is a whole family of ellipses, of major diameter $2A$ and minor diameter $2B$ ($B < A$), describable within the circle. Figure 5 shows a circle, of centre O, and one of the ellipses associated with it in this way. It is common to refer to the circle as the *auxiliary circle* of the ellipse (in the situation illustrated in the figure). If X'OX is the major diameter of the ellipse, and if QPN, drawn perpendicular to X'OX, inter-sects the circle and the ellipse in Q and P, respectively, then P and Q are said to be *corresponding points* in the ellipse and its auxiliary circle.

Let us consider the motion of a particle (in an anticlockwise sense) in the ellipse. If P represents the instantaneous position of the particle, Q is the

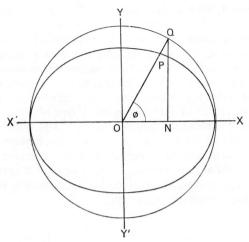

Figure 5

instantaneous position in the diagram of the corresponding point in the auxiliary circle. Obviously, we may describe the motion in terms of the time variation of the position of Q. In respect of O as origin and X'OX as reference direction, the polar coordinates of Q are (A, ϕ). If the Cartesian coordinates of P are (x, y), obviously $x = A \cos \phi$, and since the equation of the ellipse is

$$\frac{x^2}{A^2} + \frac{y^2}{B^2} = 1,$$

with respect to its principal axes, in respect of the situation illustrated $y = B \sin \phi$. Our immediate object is to identify that type of elliptical motion for which uniform motion in a circle represents the 'degenerate case'. Clearly, this can be none other than the motion described by the equation

$$\phi = \omega t. \qquad \qquad \textbf{4.28}$$

When $B \to A$, and the ellipse in Figure 5 becomes indistinguishable from its auxiliary circle, the angle ϕ of this figure is the exact counterpart of the angle θ of Figure 4, and equation **4.28** becomes identical with equation **4.26**. When equation **4.28** applies, then

$$x = A \cos \omega t, \qquad \qquad \textbf{4.29}$$
$$y = B \sin \omega t = B \cos(\omega t - \tfrac{1}{2}\pi).$$

For this special type of elliptical motion (motion for which the corresponding point describes the auxiliary circle with constant speed), we conclude that a valid alternative description is in terms of superposed simple harmonic motions of the same period $(2\pi/\omega)$, different amplitudes (A and B), constant phase difference $\tfrac{1}{2}\pi$, along the principal axes of the ellipse, and each centred at

61 Linear Motion: Speed, Velocity and Acceleration

the origin. As we should expect, when $B \to A$, equations 4.29 become identical with equations 4.27 which refer to circular motion with constant speed.

In respect of uniform circular motion we drew attention to the obvious fact that, for purposes of kinematical description, any pair of mutually perpendicular diameters is indistinguishable from any other pair, so that equations 4.27 are equally valid in relation to any choice of rectangular axes through the centre of the circle (provided that the time-zero is suitably chosen). Clearly, the situation is different in respect of the corresponding type of elliptical motion which we are now considering. Equations 4.29 are valid only for resolution along the principal axes of the ellipse. It may be shown, however – and the demonstration is left as an exercise for the reader – that if an arbitrary set of axes is taken, inclined (in an anticlockwise sense) at an angle θ to the principal axes ($-\frac{1}{4}\pi \leqslant \theta \leqslant \frac{1}{4}\pi$), then our elliptical motion is equivalent to the superposed simple harmonic motions represented by

$$
\begin{aligned}
x' &= \alpha \cos \omega t \\
y' &= \beta \cos(\omega t - \delta),
\end{aligned}
\qquad \textbf{4.30}
$$

(x', y') being the coordinates of the moving particle with respect to these axes, and the time-zero being appropriately chosen. In equations 4.30

$$
\begin{aligned}
\alpha^2 &= A^2 \cos^2 \theta + B^2 \sin^2 \theta, \\
\beta^2 &= A^2 \sin^2 \theta + B^2 \cos^2 \theta,
\end{aligned}
\qquad \textbf{4.31}
$$

$$
\text{and} \quad \tan \delta = -\frac{2AB}{(A^2 - B^2)\sin 2\theta} \quad (0 \leqslant \delta \leqslant \pi).
$$

The converse of the proposition that we have just stated is regularly quoted in textbooks on kinematics in some such terms as, 'The resultant of two simple harmonic motions of the same period, about the same origin and in directions mutually at right angles, is, in general, elliptical motion, whatever the relative amplitudes and phases of the components.' From all that has been said in our treatment of the case, it will be realized that this statement, though true, is incomplete. The resulting elliptical motion, under the conditions specified, is necessarily of the special type determined by the supplementary condition that the corresponding motion in the auxiliary circle shall be uniform motion.

In respect of equations 4.31, we have one final comment here. When $\theta = \pm \frac{1}{4}\pi$, that is when axes are taken which bisect the angles between the principal axes of the ellipse, then

$$
\alpha^2 = \beta^2 = \tfrac{1}{2}(A^2 + B^2).
$$

For this choice of axes, therefore, this special type of elliptical motion is resolvable into component simple harmonic motions of equal amplitude, whatever the eccentricity of the ellipse.

We conclude this subsection by reverting to strictly one-dimensional considerations. We consider, formally, various examples of the superposition of

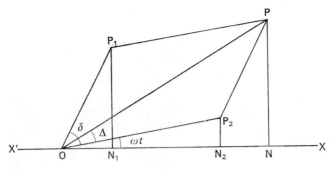

Figure 6

linear simple harmonic motions having a common origin. Suppose that we
have two such motions represented by the displacement–time equations

$$x_1 = A_1 \cos(\omega t + \delta),$$
$$x_2 = A_2 \cos \omega t. \qquad\qquad\qquad\qquad\qquad\qquad\qquad\qquad \textbf{4.32}$$

These motions are of the same period, $2\pi/\omega$, but different amplitudes, and the
former leads the latter by a constant, but arbitrary, phase angle δ. When we
say that they are superposed (or 'combined'), we imply that the actual
('resultant') displacement of the particle is at any instant given by the sum of
the 'component displacements' at that instant: $x = x_1 + x_2$. Then the displacement–time equation for the resultant displacement is

$$x = A_1 \cos(\omega t + \delta) + A_2 \cos \omega t.$$

On development, we obtain

$$x = (A_1 \cos \delta + A_2)\cos \omega t - A_1 \sin \delta \sin \omega t,$$
or $\qquad\quad x = A \cos(\omega t + \Delta),$ $\qquad\qquad\qquad\qquad\qquad$ **4.33**
where $\qquad A^2 = A_1^2 + A_2^2 + 2A_1 A_2 \cos \delta,$ $\qquad\qquad\qquad$ **4.34**

and $\quad \tan \Delta = \dfrac{A_1 \sin \delta}{A_1 \cos \delta + A_2}.$ $\qquad\qquad\qquad\qquad$ **4.35**

We conclude that the resultant of the two linear simple harmonic motions
described by equations **4.32** is the single linear simple harmonic motion
described by equation **4.33**. Self-evidently, this resultant has the same period
as its components. Its amplitude is given, in terms of the amplitudes and
relative phases of the component motions, by equation **4.34** – and its phase, in
terms of the same parameters, is given by equation **4.35**.

The results that we have just obtained may be derived alternatively, and
possibly in a more illuminating manner, by a geometrical construction based
on the original definition of linear simple harmonic motion with which we
introduced our discussion (p. 57). Let $O P_2$ (Figure 6) represent the radius of
the circle in terms of the motion in which, as in our previous use of Figure 4,

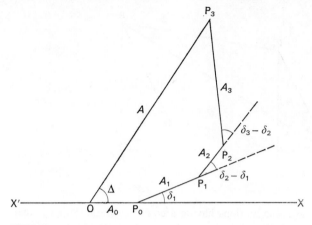

Figure 7

we define the simple harmonic motion $x_2 = A_2 \cos \omega t$ which takes place along
X'OX. Similarly, let OP_1 represent, for the same time, the radius of the
circle by which the motion $x_1 = A_1 \cos(\omega t + \delta)$ is represented. Obviously,
$\angle P_2\, OP_1 = \delta$, as indicated in the figure; also, in representational units,
$OP_1 = A_1$, $OP_2 = A_2$, $ON_1 = x_1$, $ON_2 = x_2$. Let us complete the parallelogram $P_1\, OP_2\, P$ and draw the diagonal OP. Let N be the foot of the perpendicular drawn from P to X'OX. Then, from the geometry of the figure,
$N_2\, N = ON_1$, and we have

$$ON = x_1 + x_2.$$

We conclude that, at all times (the diagram being drawn for an arbitrary time
t), the projection of OP on X'OX is equal to $x_1 + x_2$ – that is to the resultant
displacement of our problem. The resultant motion, then, is in fact simple
harmonic of the same period as that of its components, and on the scale of the
diagram the amplitude of this motion is OP. Obviously, also, $\angle P_2\, OP$ is the
phase angle Δ by which the resultant leads the lagging component. These
conclusions being accepted, equations 4.34 and 4.35 follow directly from the
geometry of the parallelogram.

It is a special merit of the graphical method that we have just described that
it may be extended very simply to deal with more complicated situations than
those that we have hitherto discussed. Suppose that we have any number of
component simple harmonic motions, of the same period, and about the same
origin in a given straight line. There will be one, of amplitude A_0, say, with
respect to which the relative phases of all the others may be represented by
positive phase angles δ_r ($0 \leqslant \delta_r \leqslant 2\pi$). Suppose that the amplitudes of these
motions are A_1, A_2, ... in the order in which the phase angle δ_r ($r = 1, 2, ...$)
increases. Then we may construct an *amplitude–phase diagram* such as that
illustrated in Figure 7, and it will be clear on the basis of our previous result

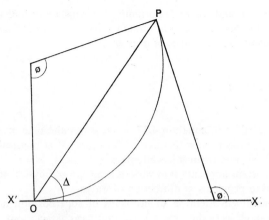

Figure 8

that the amplitude of the resultant motion will be given by the length of the line in the diagram (OP_3 in Figure 7) which completes the *amplitude polygon*. Furthermore, the phase of the resultant will likewise be given by the inclination of this line (labelled A in Figure 7) to the line which represents the last-lagging component A_0 in the diagram. (It will further be realized that in ordering the components in relation to relative phase, for the purposes of this construction, we have acted merely for sake of convenience: we have not imposed a necessary condition conferring validity on the method – the amplitude polygon might equally well have been drawn taking the components in any order, the essential results would have been unchanged.)

There is a particular case of composition of multiple motions that may be treated very directly by the amplitude–phase diagram. Suppose that, instead of a finite number of components, we have an infinite number of infinitesimal components 'uniformly distributed in phase' between limits, $0 \leqslant \delta \leqslant \phi$. By this statement we imply that the effective amplitude of the component simple harmonic motions for which the phase angle is in the range between δ and $\delta + d\delta$ is proportional to $d\delta$. We may represent this infinitesimal amplitude by $\alpha \, d\delta$, so that, if all the component motions had had the same phase, the amplitude of the resultant (A_0) would have been $\alpha\phi$. In the case specified, suppose that the amplitude of the resultant simple harmonic motion is A, and that its relative phase angle is Δ. We construct the amplitude–phase diagram and we obtain Figure 8. The amplitude polygon in this case reduces to the arc of a circle of radius α, subtending an angle ϕ at the centre. Clearly, in terms of the geometry of the figure, we have

$$A = 2\alpha \sin \tfrac{1}{2}\phi = A_0 \frac{\sin \tfrac{1}{2}\phi}{\tfrac{1}{2}\phi}, \qquad\qquad 4.36$$

$$\Delta = \tfrac{1}{2}\phi. \qquad\qquad 4.37$$

According to equations 4.36, $A \to A_0$ when $\phi \to 0$ (as, indeed, we have already stated in definition of A_0), and, more importantly, for constant α and increasing ϕ, we obtain a series of (equal) maximum values for A of absolute magnitude 2α for (equally spaced) values of ϕ given by

$$\phi = (2n+1)\pi,$$

also a series of zero values of A for

$$\phi = 2n\pi,$$

n being an integer. According to equation **4.37**, when A has alternate maximum values, effectively $\Delta = \frac{1}{2}\pi$ or $\frac{3}{2}\pi$ (a value of Δ of $2\pi + \Delta'$ is physically equivalent to the value Δ'), and when A has alternate zero values, $\Delta = \pi$ or 0. These results have important application in various branches of physics, and particularly in relation to problems of diffraction of waves.

Hitherto we have confined attention to problems of superposition and resolution in which a single periodic time τ ($\equiv 2\pi/\omega$) has characterized all the simple harmonic motions involved. As a last example of the superposition of linear simple harmonic motions, we now consider the case of two motions of which the periods are very nearly, but not quite, the same. Let the frequencies of the two motions be f_1 and f_2 ($f_1 > f_2$), and, with a suitable choice of time zero, let the motions be represented by the equations (see p. 58)

$$x_1 = A_1 \cos 2\pi f_1 t,$$
$$x_2 = A_2 \cos 2\pi f_2 t.$$

Let us write $f_1 - f_2 = f_b$.

According to our particular assumptions, f_b is a positive quantity, and $f_b \ll f_2$. On this basis, the equations representing the component motions become

$$x_1 = A_1 \cos 2\pi (f_2 t + f_b t),$$
$$x_2 = A_2 \cos 2\pi f_2 t. \tag{4.38}$$

We note that the second term in the expression for the phase angle of the first motion varies very little indeed during one complete period $1/f_2$ of the second motion; except for the effect of this term the periods of the two motions are the same. Formally, we may compare equations **4.38** with equations **4.32**. Equations **4.32** refer to component simple harmonic motions of the same period and of constant phase difference δ; as a result of our comparison we may say that equations **4.38** refer to component motions of the same period with slowly varying phase difference represented by the time-dependent angle $2\pi f_b t$. If this is a valid statement of the situation, then we may take over the results contained in equations **4.34** and **4.35** and obtain, in respect of this particular case,

$$A^2 = A_1^2 + A_2^2 + 2A_1 A_2 \cos 2\pi f_b t, \tag{4.39}$$

$$\sin \Delta = \frac{A_1}{A} \sin 2\pi f_b t; \tag{4.40}$$

(we have transformed equation **4.35** in such a way as to obtain an expression in which the denominator is essentially positive throughout). The interpretation of equations **4.39** and **4.40** is clearly as follows: the amplitude of the resultant motion (of frequency f_2) varies harmonically ('sinusoidally') with time, between limits $A_1 + A_2$ and $A_1 - A_2$ with frequency $f_b (\equiv f_1 - f_2)$; as the amplitude passes through its maximum and minimum values, the sign of the phase angle Δ changes from negative to positive and positive to negative, alternately. In respect of real situations, equations **4.39** and **4.40** describe the phenomenon of *beats*: the difference frequency f_b is the *beat frequency* – the resultant amplitude rises and falls sinusoidally with a frequency equal to the difference of the frequencies of the two simple harmonic motions which are superposed.

4.3   Motion involving rotation

4.3.1   *General considerations*

We have already quoted Maxwell (p. 19) as saying, 'But we cannot treat [bodies, whether large or small,] as material particles when we investigate their rotation.' As, therefore, we entitled the last section 'linear motion' – and thereby accepted any topic in particle kinematics as within our purview – we now understand the title of the present section to exclude the further consideration of particles, leaving the way open for the discussion of selected topics in the kinematics of extended bodies.

The concept of a point-particle represents the ultimate in idealization of the entities of the real world: this having been said, within the subject matter of kinematics there is no cause for its further elucidation. The concept of an extended body is less precise. We shall devote the next chapter to the consideration of some of the macroscopic properties of real bodies, and of their idealization in the 'rigid body' concept which is used extensively in the (approximate) solution of dynamical problems. That is the appropriate order of development. Here it is our purpose to restrict consideration to the kinematical description exclusively, as a preparatory exercise in 'background mathematics'. Under this limitation we may postpone the use of the categories 'real body' and 'rigid body', for the time being, and use instead the purely formal category 'figure in Euclidean space'. We shall be concerned, then, in this section with the description of the motion of figures in Euclidean space – accepting the axiom of transferability (p. 34).

4.3.2   *Two-dimensional situations*

Let us consider, first, motion in space of two dimensions. Our generalized figure in this case is a plane figure bounded by a perimeter, polygonal or curvilinear as the case may be. The size and shape of the figure being determined, its location in the two-dimensional space may be completely specified

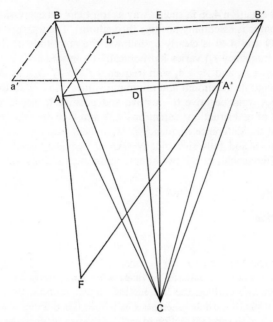

Figure 9

in terms of the positions of any two points fixed in the figure. For the purposes of formal discussion, therefore, the kinematical problem reduces to that of the motion of a finite rectilinear filament in two dimensions.

Suppose, then, that AB (Figure 9) represents the 'initial' and A'B' the 'final' position of such a representational filament in respect of a small displacement. We wish to consider this displacement in relation to the independent aspects of 'pure translation' and 'pure rotation'. By 'pure translation', in this connexion, we imply a change of position in which the vector displacement of every point in the filament is the same – the filament moves between initial and final positions which are parallel one to the other: by 'pure rotation' we imply movement about a point fixed in our two-dimensional frame of reference – every point in the filament moves along an arc of a circle having this point as centre. It is a matter almost of definition that the component of pure rotation, in the displacement represented in the figure, is uniquely determined. If BA and B'A', when produced, meet in F, ∠ BFB' is the angle between the initial and final directions of the filament, and a rotation of this amount must certainly be combined – or so we should assume – with any motion of pure translation by which the final displacement is to be achieved. With respect to the component of pure translation the conclusion is less obvious: it is our immediate object to show, by means of examples, that this component can take on an infinity of values (including zero) depending upon our choice of centre of rotation for the other component.

Suppose, then, that we choose A – the point in space initially occupied by one end of the filament – as the centre of rotation. The component of pure rotation transfers the filament from AB to Ab' (Ab' being parallel to A'B'). The component of pure translation necessary to complete the transfer to A'B' is then $\overrightarrow{AA'}$. Obviously, the magnitude of the rotation is given by $\angle BFB'$. Alternatively, suppose that B – the point in space initially occupied by the other end of the filament – is chosen as rotation centre. Rotation about B through an angle equal to $\angle BFB'$ transfers the filament from AB to a'B. Thereafter, a component of pure translation $\overrightarrow{BB'}$ is required to complete the specified displacement. Comparing these two examples, clearly, $\overrightarrow{AA'} \neq \overrightarrow{BB'}$. Finally, consider C, the point of intersection of DC and EC, the perpendicular bisectors of AA' and BB', respectively, as centre of rotation. $\triangle ABC$ and $\triangle A'B'C$ are congruent, since

AB = A'B',

by the axiom of transferability,

and   BC = B'C,
      CA = CA',

as a result of our construction for C. Obviously, pure rotation, about C as centre, through $\angle BCB'$ ($= \angle ACA'$) transfers $\triangle ABC$ into congruence with $\triangle A'B'C$, and, in particular, transfers AB into coincidence with A'B'. In C, then, we have identified a centre of pure rotation such that the specified displacement AB to A'B' is effected solely by rotation about this centre, without the necessity for superposed translation. In order to complete our discussion, and confirm our original semi-intuitive conclusion, we have merely to show that the magnitude of the rotation about this unique centre is of the expected value, namely, that $\angle BCB' = \angle BFB'$. In order to do this we note that $\angle BCB'$ is the supplement of $\angle CBB' + \angle CB'B$, whereas $\angle BFB'$ is the supplement of $\angle FBB' + \angle FB'B$.

Moreover, $\angle FBB' = \angle CBB' + \angle ABC$,
            $\angle FB'B = \angle CB'B - \angle A'B'C$.

Now, as we have already shown,

$\angle ABC = \angle A'B'C$.

Thus $\angle FBB' + \angle FB'B = \angle CBB' + \angle CB'B$,

and, in consequence,

$\angle BFB' = \angle BCB'$,

as we set out to prove. Simple rotation, about the unique centre C, of magnitude equal to $\angle BFB'$, transfers the representational filament of our discussion (or any other filament in terms of which the motion of the two-dimensional figure of our problem is alternatively described) from its initial

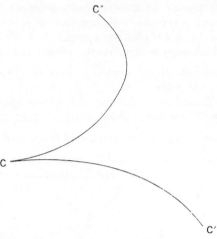

Figure 10

to its final position, in respect of the arbitrary small displacement that we have postulated.

When we are concerned with the continuous motion of a two-dimensional figure, we proceed to the limit, as in the one-dimensional case (p. 49), and consider an infinite sequence of infinitesimal displacements as taking place consecutively. In this way we are led to the concept of an instantaneous centre of simple rotation of the figure – and a unique instantaneous angular velocity of rotation about that centre – but we are not, in general, led to the definition of a unique instantaneous velocity of translation. As the motion progresses, in general the position in (two-dimensional) space of the instantaneous centre of simple rotation changes: in such a case we refer to the locus of this point as the *space centrode* of the motion. At each instant in time, some point within the figure – or in a frame of reference moving with the figure – is instantaneously coincident with the centre of simple rotation in space: the locus of this point 'belonging to' the figure, which is instantaneously at rest in space, is referred to as the *body centrode* in the frame of reference moving with the figure. A little consideration will show that, in respect of any finite stretch of continuous motion of a plane figure, the total curvilinear length of the body centrode of the motion is precisely the same as that of the space centrode: the former curve, in fact, 'rolls without slipping' on the latter. We may establish this result simply in terms of Figure 10. CC′ is a finite section of the space centrode of the motion of a plane figure in the plane of the diagram. CC″ is the position occupied in this plane, at the 'initial' instant of time, by the corresponding section of the body centrode. At the initial instant, the figure is rotating about C. At the 'final' instant, it is rotating about C′, and the point in the figure which was initially at C″ has by this stage moved to C′. For intermediate stages in the motion we can identify corresponding points in

CC′ and CC″ and, because at each stage the motion is one of simple rotation about the appropriate point in CC′ (which by definition is a curve fixed in space), the progress of the motion can evidently be described (as we have asserted) by the 'rolling without slipping' of CC″ on CC′. Corresponding sections of these curves are necessarily of the same length: each point in CC″ ultimately approaches into coincidence with, and then recedes from, the corresponding point in CC′ along an element of path which is perpendicular to the tangent to that curve at the point concerned.

The essential existence theorem regarding the instantaneous centre of simple rotation in two-dimensional motion – and the theorem regarding the properties of the centrodes that we have just established – are both due to Michel Chasles (1793–1880). Many textbooks develop the arguments that lead to these conclusions in relation to the 'plane motion' of a three-dimensional figure, rather than in respect of the plane motion of a plane figure, as we have done. There is no essential difference between these two formulations of the problem. If, in respect of a three-dimensional figure, we understand by 'plane motion' motion in which each point in the figure moves in a plane parallel to a fixed reference plane in three-dimensional space, then no new feature is involved. Our three-dimensional figure may be regarded as built up of an infinite stack of mutually related plane figures, stratified parallel to the reference plane in question, and the only significant change in our verbal description of the situation will be the replacement of the term 'instantaneous centre of simple rotation' by 'instantaneous axis of simple rotation'. All such axes of rotation will of necessity be perpendicular to the specified reference plane, and the definitions of the centrodes may be retained as the loci of points moving in that plane.

We have just established the general proposition that any arbitrary (small) displacement of a two-dimensional figure in its own plane may be regarded as effected by simple rotation about a unique centre. Accepting the converse of this proposition, that an arbitrary (small) rotation of a two-dimensional figure about any point in its plane may be regarded as effecting a small displacement of the figure which is of full generality, we shall find it useful to discuss the resultant displacement which ensues when two or more such rotations are superposed.

Let us suppose, therefore, that a two-dimensional figure is instantaneously subject to simultaneous rotations, with angular velocities ω_1 and ω_2, respectively, about two points P_1 and P_2 in its own plane. Let the rectangular coordinates of these points, with respect to fixed reference axes, be (x_1, y_1) and (x_2, y_2). Consider an infinitesimal time interval Δt, during which the coordinates of an arbitrary point in the figure (with respect to the same axes) change from (x, y) to $(x + \Delta x, y + \Delta y)$. Then, if the rotation about P_1 alone had been effective, we should have had, in obvious notation,

$$(\Delta x)_1 = -(y - y_1)\omega_1 \, \Delta t,$$
$$(\Delta y)_1 = (x - x_1)\omega_1 \, \Delta t,$$

4.41

and a similar pair of equations would represent the other possibility that only rotation about P_2 was in question. Under the conditions of superposition that we have postulated (compare p. 63),

$$\Delta x = (\Delta x)_1 + (\Delta x)_2,$$
$$\Delta y = (\Delta y)_1 + (\Delta y)_2.$$

In these circumstances,

$$\Delta x = -(y - y_1)\omega_1 \Delta t - (y - y_2)\omega_2 \Delta t,$$
$$\Delta y = (x - x_1)\omega_1 \Delta t + (x - x_2)\omega_2 \Delta t. \tag{4.42}$$

Now, equations 4.42 may be written in the form

$$\Delta x = -(y - \bar{y})\omega \Delta t,$$
$$\Delta y = (x - \bar{x})\omega \Delta t, \tag{4.43}$$

if $\quad \omega = \omega_1 + \omega_2,$ $\tag{4.44}$

and $\quad \bar{x} = \dfrac{x_1 \omega_1 + x_2 \omega_2}{\omega_1 + \omega_2},$

$$\bar{y} = \dfrac{y_1 \omega_1 + y_2 \omega_2}{\omega_1 + \omega_2}. \tag{4.45}$$

Equations 4.44 and 4.45 defining, respectively, an angular velocity and the rectangular coordinates of a point in the fixed frame of reference (in terms of the angular velocities and rectangular coordinates 'given' in the problem), it is an obvious conclusion, from the comparison of equations 4.43 with equations 4.41, that the resultant displacement we are seeking is describable in terms of the simple rotation of the figure about the point $(\bar{x}, \bar{y})$ with angular velocity ω. We have shown, in fact, that the resultant of two simultaneous simple rotations (in this two-dimensional situation) is instantaneously also a simple rotation (in general), of angular velocity equal to the (algebraic) sum of the angular velocities of the components – and we have obtained expressions (equations 4.45) which identify its centre. We note that this centre necessarily lies in the line joining the centres of the component rotations, dividing this line in the inverse ratio of the corresponding angular velocities: from equations 4.45

$$\frac{\bar{x} - x_1}{x_2 - \bar{x}} = \frac{\bar{y} - y_1}{y_2 - \bar{y}} = \frac{\omega_2}{\omega_1}. \tag{4.46}$$

Obviously, under the same limiting conditions as we have already imposed, this last result may be extended to include the case of the composition of any number of component rotations. In that case, self-evidently, equations 4.44 and 4.45 take the form

$$\omega = \sum_r \omega_r, \qquad\qquad\qquad\qquad\qquad \textbf{4.47}$$

$$\bar{x} = \frac{\sum_r x_r \omega_r}{\sum_r \omega_r},$$

$$\bar{y} = \frac{\sum_r y_r \omega_r}{\sum_r \omega_r}. \qquad\qquad\qquad\qquad\qquad \textbf{4.48}$$

We have said that, instantaneously, the resultant of two simultaneous simple rotations is, in general, also a simple rotation. A special case in which this conclusion is more conveniently expressed in other terms is that in which the angular velocities of the components are equal and oppositely directed. When $\omega_2 \to -\omega_1$, then (see equations **4.43** to **4.45**)

$$\omega \to 0, \qquad \bar{x} \to \bar{y} \to \infty,$$

and $\quad \Delta x \to (y_1 - y_2)\omega_1 \Delta t,$
$\qquad\quad \Delta y \to (x_2 - x_1)\omega_1 \Delta t.$

We note that the expressions for Δx and Δy, in this case, no longer contain x and y, the coordinates of the point in the figure to which the displacement refers. Every point in the figure, then, experiences the same (vector) displacement: the resultant of the two simple rotations, in this special case, is a pure translation (or a simple rotation about 'a point of infinity').

We make one final comment on our recent discussions. Throughout our argument, by the explicit introduction of the infinitesimal Δt, and by restricting consideration to the 'instantaneous' situation, we have retained generality, and have expressed our conclusions in terms which are applicable to the unconstrained two-dimensional motion of a figure – that is to motion for which the space centrode is an arbitrary curve. In fact, our most recent conclusions have their simplest application to constrained motion. If, for example, a two-dimensional figure simply rotates continuously in its own plane about a fixed point (so that the body centrode, also, degenerates into a single point), then this motion may be resolved into continuous simple rotations about any two other points in the plane, provided that equations **4.44** and **4.45** are satisfied. Obviously, a corresponding statement may be made concerning the composition of such simple rotations in this case.

4.3.3   *Three-dimensional situations*

In order to specify the location in three-dimensional space of a three-dimensional figure, it is necessary (and sufficient) to assign positions in space to three points which are non-collinear in the figure. In the most general motion of such a figure the positions in space of all three reference points change continuously. A less general type of motion (to which we shall devote more attention here) is that in which one point in the figure remains fixed in space uniquely. In that case it is only necessary to assign positions in space to two other points in the figure in order to specify its location. Moreover, there is no loss in generality if these two points are equidistant from the fixed point (provided that the three points are not collinear).

Let us, therefore, examine the possibilities of motion of a three-dimensional figure having one point fixed by use of two reference points fixed in the figure and equidistant from O, the point in the figure which remains fixed in space. Clearly, whatever the motion, under this constraint, these reference points remain on the surface of a sphere centred in O – and their separation, measured along a great circle of this sphere, remains constant. (A *great circle* is determined by the intersection of any diametral plane with a spherical surface.) On the basis of this mode of representation an arbitrary small displacement of the figure, with O fixed, is described in terms of the displacement, on the surface of this reference sphere, of a great-circle filament of constant length from an 'initial' position AB to a 'final' position A'B'. If we regard Figure 9 (p. 68) as being drawn on the spherical reference surface, all the lines in the diagram now being great-circle arcs, we may interpret the point C in the diagram, without further ado, as the point on the reference sphere about the normal through which simple rotation will transfer the representative great-circle filament from the initial position AB to the final position A'B'. This interpretation establishes the proposition that any small displacement of a three-dimensional figure having one point fixed in space may be regarded as a simple rotation about a unique axis through the fixed point. The most general infinitesimal displacement, then, under this constraint, is an infinitesimal rotation about an arbitrary axis through the fixed point. This result was first established by Leonhard Euler (1707–83) in 1776.

We have obtained our last result by taking over the geometrical formalism appropriate to the case of the unconstrained motion of a two-dimensional figure in its own plane and applying it to the motion of a three-dimensional figure having one point fixed in space. This is a valid application, fundamentally, because the figures concerned in the two situations have the same number of degrees of freedom (see section 4.4). Let us now transfer another result from the two-dimensional to the three-dimensional situation, re-interpreting it appropriately. We note that the three-dimensional analogue of rotation about an arbitrary point in the fixed two-dimensional reference plane of the former situation is rotation about an arbitrary axis through the centre of the fixed reference sphere of the latter situation. We may therefore transfer the result that we obtained regarding the composition of simultaneous rotations instantaneously effective about different points in the two-dimensional case, and apply it to the composition of simultaneous rotations instantaneously effective about different axes through the fixed point in three dimensions. We conclude, at once, that if the instantaneous motion of a three-dimensional figure with one point fixed in space, is rotation about a specified axis through that point with angular velocity ω, then this motion may be resolved into

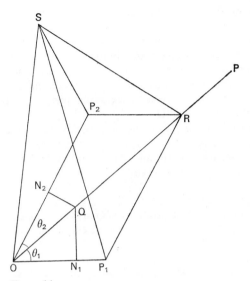

Figure 11

instantaneous rotations of appropriate velocities about any two other axes passing through the fixed point and lying in the same plane with the specified axis of instantaneous rotation. Conversely, instantaneous rotations about any two axes fixed in space and intersecting in the point in the body which is fixed, compound into a simple instantaneous rotation about a unique axis which likewise passes through the fixed point and lies in the plane of the other two. Obviously, these statements regarding composition and resolution of instantaneous rotations can be extended without difficulty to include the case of multiple components, as was done in the previous discussion.

It will be instructive to reconsider the problem of the composition of instantaneous rotations independently, by direct analysis: in so doing we shall hope to substantiate, and render quantitative, the merely qualitative statements that we have just made on the basis of analogy. Suppose, then, that O (Figure 11) represents the point in a three-dimensional figure that is permanently at rest in space. Let the vectors $\overrightarrow{OP_1}$, $\overrightarrow{OP_2}$ represent (see p. 57) the instantaneous magnitudes and axial directions of the angular velocities ω_1 and ω_2 of which we wish to identify the instantaneous resultant. Let us consider an infinitesimal time interval Δt. If the motion about OP_1 alone were effective, any point in our three-dimensional figure instantaneously lying in the plane of the diagram and at a distance r_1 from OP_1 would be displaced, in a direction perpendicular to the diagram, through a distance $r_1 \omega_1 \Delta t$. This displacement would (conventionally) be towards the reader for points lying 'above' OP_1 and away from the reader for points lying 'below' OP_1. Similarly, if the motion about OP_2 alone were effective, the corresponding displacement would be $r_2\,\omega_2\,\Delta t$, towards the reader for points lying above OP_2 and *vice versa*. If the motions about OP_1 and OP_2 were effective together, points lying above OP_2, or below OP_1, would be subject to component displacements in the same sense – towards the reader in the one case, and away from the reader in the other – and points lying between the two axes would be subject to component displacements of opposite sense. We wish to show that, in the last case, these component displacements cancel uniquely for points lying in a particular straight line through O.

Let us suppose that the straight line OP (Figure 11) satisfies the condition that we have just imposed. Let Q be any point in this line and let QN_1, QN_2 be drawn so as to be perpendicular to OP_1 and OP_2, respectively. Then, in respect of Q, our determining condition is given by

$$\omega_1\,QN_1 = \omega_2\,QN_2,$$

and this condition must be satisfied for all positions of Q in OP. According to the figure, $QN_1 = OQ\sin\theta_1$, $QN_2 = OQ\sin\theta_2$; thus the condition becomes

$$\omega_1\sin\theta_1 = \omega_2\sin\theta_2. \tag{4.49}$$

Equation 4.49 does not explicitly involve the position of Q: taken together with the equation

$$\theta_1 + \theta_2 = \theta, \tag{4.50}$$

($\angle P_1\,OP_2 = \theta$), it enables a unique direction, through O and within the sector $P_1\,OP_2$, to be specified completely. Clearly, through equations 4.49 and 4.50, we have established the limited conclusion that we wished to draw. We may now paraphrase our original statement of that conclusion as follows: when a three-dimensional figure having one point fixed is simultaneously subject to rotations about two axes passing through the fixed point there is a particular line in the figure, also passing through the fixed point and lying in the plane of these axes, which is instantaneously at rest in space.

This conclusion having been reached, we can hardly imagine otherwise than that the instantaneous motion of the figure is simple rotation about the instantaneously fixed line that we have identified, but we have not as yet formally established this result, at least in the sense of deriving an expression giving the instantaneous angular velocity of rotation about this line. This we must now do. Let us consider again the infinitesimal displacement of S, any point in the plane of the axes of ω_1 and ω_2, during the time interval Δt. We have already shown that the resultant displacement of S (Figure 11) is along the normal to this plane, and we have discussed the way in which the sense of the resultant displacement depends upon the position of S in this plane. For positions above OP_2 and below OP_1 in the diagram our previous results are consistent with the assumption that these displacements can be ascribed to a single positive rotation about OP.

In absolute magnitude the resultant displacement of S is given by

$$(r_1\,\omega_1 + r_2\,\omega_2)\,\Delta t$$

in our previous notation. Now, on the scale of the diagram, the measure of ω_1 is represented by OP_1, and that of ω_2 by OP_2. On the corresponding scale, therefore, the measure of $r_1 \omega_1$ is represented by twice the area of $\triangle SOP_1$, and the measure of $r_2 \omega_2$ by twice the area of $\triangle SOP_2$. These triangles have been drawn in the diagram. If we can identify a point R in OP such that, in respect of area,

$$\triangle SOR = \triangle SOP_1 + \triangle SOP_2, \qquad \textbf{4.51}$$

obviously we can interpret the length of OR, in the scale of the diagram, as the measure of ω, the angular velocity about OP, which, effective for a time Δt, produces the same displacement of the arbitrary point S as is produced by the component rotations that we are considering. If R (Figure 11) is in fact the point for which equation **4.51** is valid, then, because

$$\triangle SOR = \triangle SOP_2 + \triangle ORP_2 + \triangle RSP_2,$$

we must necessarily have

$$\triangle SOP_1 = \triangle ORP_2 + \triangle RSP_2. \qquad \textbf{4.52}$$

Equation **4.52** is satisfied if $P_2 R$ is parallel and equal to OP_1. If $P_2 R$ is parallel to OP_1, $\angle P_2 RO = \theta_1$, and we have

$$P_2 R \,.\, \sin \theta_1 = OP_2 \,.\, \sin \theta_2. \qquad \textbf{4.53}$$

However, because we have drawn the diagram 'to scale' (see above),

$$OP_2 \,.\, \omega_1 = OP_1 \,.\, \omega_2. \qquad \textbf{4.54}$$

Clearly, equations **4.53** and **4.54** are inconsistent with equation **4.49** unless

$$P_2 R = OP_1.$$

On the basis of our earlier construction, then, fixing the direction of OP, if $P_2 R$ is parallel to OP_1, it is also equal in length to it.

We have developed the construction for the vector representing the resultant rotational velocity in this case by first defining the direction of the vector (by the trigonometrical equations **4.49** and **4.50**) and then defining its length by a further geometrical construction, $P_2 R \| OP_1$ (Figure 11). Our last result indicates that these two procedures might in fact have been combined: OR is simply the diagonal through O of the parallelogram of which OP_1 and OP_2 are adjacent sides. Indeed, the basic parallelogram construction, which we discussed first in relation to linear displacement vectors (p. 23), appears in this instance to be equally valid when angular velocities of rotation (axial vectors) are concerned.

Let us return to the analogy between unconstrained two-dimensional motion and constrained (one-point-fixed) motion in three dimensions. It will be recalled that we concluded (p. 74) that there exists a valid analogy between centres of rotation in the former situation and axes of rotation through the fixed point in the latter. In the former situation, in respect of any continuous motion of a plane figure we defined the space centrode and the body centrode of the motion as the loci, in space and in the moving figure, respectively, of the successive positions occupied by the instantaneous centre of simple rotation (p. 70) – and we showed that the relation between these two loci is that the latter rolls on the former, without slipping, as the motion proceeds (p. 71). Obviously, in general, the 'locus' of a line which, passing through a fixed point, changes its direction in space continuously, is a continuous surface having the fixed point as apex. We shall refer to such a surface, of whatever shape, as a cone. In terms of this phraseology, then, and following our analogy in relation to the continuous motion with which we are now concerned, we assert that the continuous motion of a three-dimensional figure having one point fixed in space is describable in terms of the rolling (without slipping) of one cone on another, these cones having a common apex in the fixed point. The cones themselves are generated by the motion in the frame of the moving figure, and in the fixed reference frame, respectively, of the instantaneous axis of simple rotation which characterizes the motion.

Hitherto, in this section, in so far as we have dealt with the continuous motion of extended figures, in two-dimensional and three-dimensional space, we have concentrated attention on the description of the motion in terms of the progressive change in the location of an instantaneous centre of rotation or of an instantaneous axis of simple rotation, as the case may be. Any such change (in an instantaneous angular velocity) obviously implies that the rotational motion is accelerated motion, but we have not sought to specify the nature of the relevant acceleration in any case. There is a particular class of situation in which this may be done without great complication: here we wish to consider one of these situations very briefly.

When a three-dimensional figure with one point fixed in space rotates with constant angular speed ω about an axis through that point and fixed in the figure, and this axis rotates about an axis through that point and fixed in space with constant angular velocity Ω, maintaining a constant inclination θ to the fixed axis, the motion is referred to as *precessional* motion. The simplest case of such precessional motion is obviously that for which $\theta = \frac{1}{2}\pi$. In this case the 'body cone' of instantaneous axes of rotation degenerates into a straight line and the 'space cone' degenerates into a plane in which that line rotates with angular velocity Ω about the fixed point. In these circumstances the vector diagram representing an infinitesimal change in the angular velocity of the figure (in an element of time Δt) is an elementary sector of a circle having an angle $\Omega \Delta t$ at the centre O (Figure 12). On the basis of the parallelogram construction, $\overrightarrow{P_1 P_2}$ in the diagram represents the increment of angular velocity in the time interval Δt – on the same scale as the measure of ω is represented by the length of $\overrightarrow{OP_1}$ – and, when we proceed to the limit, we have for α, the measure of the angular acceleration concerned,

$$\alpha = \omega \cdot \frac{\Omega \, \Delta t}{\Delta t},$$

4.55

or $\alpha = \Omega\omega$.

The axis about which this angular acceleration is effective is given by Figure 12 as the line through the fixed point parallel to $\overrightarrow{P_1 P_2}$: it is instantaneously at right angles both to the instantaneous position of the axis of rotation of the figure and to the axis in space about which this axis is rotating (with angular velocity Ω). This statement, and equation **4.55**, which is given here in non-vectorial form, may be compared with the corresponding statement and equation **4.20** (p. 56) which had reference to particle motion in a circle with uniform speed. The close formal relationship between the two situations becomes immediately apparent in such a comparison. Essentially, the two situations, one involving linear motion and the other angular motion, are such that formally identical applications of the parallelogram construction are relevant: necessarily, formally identical results emerge.

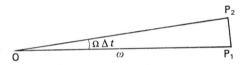

Figure 12

Our last concern in this subsection is to describe, in the simplest possible terms, the instantaneous situation obtaining when a three-dimensional figure moves without constraint in three-dimensional space. We proceed, as before, by formulating a description of the most general type of small displacement which our specification allows. We have previously shown (p. 74) that, when one point in the figure remains fixed in space, an arbitrary 'most general' displacement is one of rotation about an identifiable axis through the fixed point. Evidently, then, if this 'single point' constraint is removed, the most general displacement is one of pure translation (p. 68), of arbitrary magnitude and direction, followed by a simple rotation of arbitrary magnitude about any axis. However, this form of description is not the simplest possible, as we shall now proceed to show.

In a pure translation in three dimensions, each plane in a figure remains parallel to its initial 'direction'; in a simple rotation, each plane which is perpendicular to the axis of rotation retains its 'direction' unaltered. In the most general displacement (as we have just described it), therefore, each plane in the figure which initially was perpendicular to the direction of the ultimate axis of rotation will still be found, at the end of the two-stage process of translation followed by rotation, to be perpendicular to that axis. Let us consider three non-collinear points in such a plane, and let us denote their initial disposition in the fixed reference space by the symbol σ and their final disposition, after the two-stage process that we have described, by σ'. We are agreed (p. 73) that the initial and final locations in space of our three-dimensional figure are uniquely specified by these dispositions. Moreover, from what we have said, the transition from σ to σ' could equally well have been effected by a pure translation in the direction of the ultimate axis of rotation of our previous description followed by a pure rotation about an axis parallel to that axis and through an identifiable point in the plane of σ' (see p. 69). We have shown, then, that an arbitrary 'most general' displacement, of a three-dimensional figure under conditions of no constraint is describable as a pure translation followed by a simple rotation about an axis parallel to the direction of translation. Such a related combination of translation and rotation may be referred to as a *twist*. Basing our nomenclature on common technical usage in relation to screws in engineering practice, we may define the *pitch* of the twist in any case as the linear parameter z/θ, z being the magnitude of the axial (translational), and θ the magnitude of the angular (rotational) displacement, involved. Obviously, when we proceed to the limit of very small displacements, it becomes immaterial whether we consider the translational and rotational components of displacement as occurring consecutively or simultaneously: ultimately, then, we may describe the instantaneous state of most general motion, in the case in question, in terms of a velocity of pure translation in a specified direction and a (superposed) angular velocity of rotation about an axis having the same direction. Chasles (see p. 71) first obtained this result in 1830.

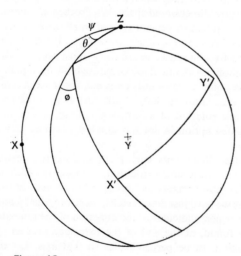

Figure 13

4.4 Degrees of freedom

In subsection 4.3.3 we made considerable use of the argument by analogy in deriving results relative to the motion of a three-dimensional figure having one point fixed in three-dimensional space from 'corresponding' results previously established in relation to the unconstrained motion of a two-dimensional figure in its own plane. We justified this procedure (p. 74) by the statement that in these two situations the moving figure is characterized by the same number of degrees of freedom. In this section we wish briefly to examine the concept of degrees of (kinematic) freedom more systematically.

We say that the number of degrees of kinematic freedom which characterizes any figure, in a specified situation, is the smallest number of independent numerical parameters (coordinates) of which the magnitudes must be known if the location of the figure is to be completely determined.

Let us consider the most complicated case that we have so far encountered, that of the three-dimensional figure subject to no constraint. We have said that it is sufficient to know the coordinates of any three non-collinear points fixed in the figure in order to determine its location (p. 73). Now, the Cartesian coordinates of three points involve nine independent magnitudes. However, when we have expressed the condition that the three points maintain their relative separations constant (that is, maintain a constant configuration – p. 19) we have three relations involving these nine magnitudes, whereby (any) three of them become redundant. Consequently, the number of degrees of freedom of the figure, in this situation, is precisely six.

We may note that it is not sufficient to know the coordinates of three collinear points in this case if the location of the figure is to be fully defined. When the points are collinear, knowledge in respect of the third point adds nothing to the information provided by the coordinates of the other two. We are left with six coordinates, and one relation expressing the constancy of separation of the two points concerned: only five independent magnitudes are known, if these two points are fixed. In such circumstances the figure is constrained so that one line in the figure remains fixed in space; it has one remaining degree of freedom – that of rotation about this fixed line. The motion of the figure may then be regarded as one dimensional – as is the motion of a point in a straight line – and in respect of this one-dimensional angular motion the simple kinematical equations 4.5, 4.6 and 4.7, which we derived for the case of the rectilinear motion of a point-particle (p. 52), may be applied with the appropriate change of symbols.

When one point in a three-dimensional figure is fixed in Euclidean space, the number of degrees of freedom of the figure is obviously reduced from six to three (specification of the three Cartesian coordinates of the point is necessary to fix its position). Because the most general displacement of the figure in such a situation is one of rotation about an axis through the fixed point (p. 74), it is often convenient to use three angular coordinates to represent the parameters to which the remaining degrees of freedom permit variation. To

81 Degrees of Freedom

this end let us establish a system of rectangular axes, OX', OY', OZ', having origin at the fixed point and fixed in the figure, and another set, OX, OY, OZ, fixed in space and having the same origin. Our aim is first to identify the direction of one of the axes belonging to the figure (say OZ') – which necessitates the specification of two angles – and then by specification of the degree of rotation about this axis to define the situation completely. We use θ to denote the angle which OZ' makes with OZ, and ψ to denote the angle which the plane ZOZ' makes with the plane ZOX, then we complete the specification in terms of ϕ, the angle which the plane $Z'OX'$ makes with the plane ZOZ'. These angles (referred to as the Eulerian angles) are illustrated in Figure 13. The diagram is drawn to represent great circles seen on (the front hemispherical surface of) a sphere centred at O. The points labelled X, Y, Z are the points of emergence through the sphere of the rectangular axes OX, OY, OZ which are fixed in space. Similarly, $X'Y'Z'$ are the points of emergence of the rectangular axes fixed in the figure. The angle θ is the angle subtended by ZZ' at O; the angles ψ and ϕ are the spherical angles indicated. Using the language of the geographer, we may say that the Eulerian angles comprise one angle of co-latitude and two angles of longitude (about separate polar axes and in relation to two different meridional planes).

Reverting now to the two-dimensional figure in two-dimensional space, as we have previously asserted (p. 68), knowledge of the positions in this space of two points belonging to the figure is sufficient to determine its position uniquely. In this case each point is characterized by two Cartesian coordinates and there is the one condition that the separation of the points is invariable: the unconstrained figure has three degrees of freedom as we have stated already (pp. 74, 81). If such a figure has one point fixed in two-dimensional space, the number of degrees of freedom is reduced to one – the only motion possible for the figure is simple rotation about that point as centre. Again, as in the case of the three-dimensional figure with an axis fixed in three-dimensional space, this motion may be regarded as one-dimensional (angular) motion.

Further reading

P. W. Bridgman, *The Logic of Modern Physics* (1927), Collier-Macmillan, 1946.

W. C. Michels, M. Correll and A. L. Patterson, *Foundations of Physics*, Van Nostrand, 1968.

K. R. Popper, *The Logic of Scientific Discovery* (2nd edn), Hutchinson, 1959.

E. F. Taylor, *Introductory Mechanics*, Wiley, 1963.

A. N. Whitehead, *An Enquiry concerning the Principles of Natural Knowledge*, Cambridge University Press, 1925.

Chapter 5
Bodies

5.1 Introductory

In this chapter we have two aims: one that can be achieved without elaborate discussion, the other necessitating more detailed consideration. Our first aim is to point the distinction between the diverse characteristics that we know bodies in the real world to possess and the simple characteristics (or lack of characteristics) which we assume them to possess when we seek to formulate an approximate description of their dynamical behaviour. Our second aim is of a different nature: it is to describe some of the experiments that were made on bodies – bodies 'falling under gravity', or in collision one with another – in the early days, before a satisfactory general method of describing their dynamical behaviour had been developed by Newton. As we have already said (p. 67), this is the appropriate place in which to introduce these considerations. In the next chapter we have to consider the Newtonian concepts of mass and force: these would not have been developed in the way that they were developed, had there not already been on record the experimental observations that we shall presently describe.

5.2 Real and ideal bodies

The word 'body' has no absolutely precise meaning in physics, though it is widely used. Broadly, it may be said to denote any 'parcel' of matter which may be identified as cohering throughout the particular period of time during which we are interested in its behaviour – in its intrinsic properties, or in its general behaviour as a component of a larger system. Thus a dust cloud in interstellar space is a body, a volume of gas enclosed within the bulb of a gas thermometer is a body, falling raindrops and globules of mercury resting on flat surfaces are bodies, so are the innumerable solid objects which surround us in our daily life, in so far as we are interested in them.

Traditionally, the subject of dynamics deals almost exclusively with the kinetic behaviour of bodies which are 'solid'. In this chapter we shall follow tradition and limit our discussion accordingly. Even so, it will be obvious from the outset that the class of objects involved is by no means a simple one. The matter of which a solid body is composed may be homogeneous or heterogeneous, and it is not unreasonable that we should include in our survey both those bodies which are solid throughout and those which are 'hollow'. In respect of intrinsic properties, also, there is a wide range of variation. In relation to compression, a lump of putty, a rubber ball and a block of steel behave very differently; in relation to more violent treatment, the properties of bodies made of steel and cast iron are in no way the same. Some bodies are magnetizable to great intensity, most are not; some bodies develop electrical polarity under compression, though this phenomenon is rare enough. All those differences, in the last analysis, are relevant to any full consideration of the mutual actions of bodies. In dynamics they are largely, if not completely, ignored.

In the subject of dynamics (kinetics) we operate almost exclusively with the concept of the *ideal rigid body*. Effectively, we take over the notion of the 'three-dimensional figure in Euclidean space', with which we were concerned in the last chapter, and we endow this figure with substance, thereby making a model of a real body. We assume that the Euclidean axiom of transferability (pp. 34, 67) applies to every motion of the body, without exception; in this way we postulate that the configuration of 'ideal particles' which constitute the body remains permanently unchanged. An ideal rigid body retains its size and shape indefinitely, through every vicissitude with which we may be concerned. Obviously, there are 'extreme' situations for the discussion of which such a model is patently inadequate; equally obviously, there is no situation of real physical interest into the discussion of which the use of the model does not introduce some approximations. Fortunately, in a large class of real situations, the limitations imposed by these approximations are of little consequence – leading to insignificant errors in prediction; if it had not been so, the subject of dynamics would almost certainly have developed very differently from the way in which it has in fact developed during the last three centuries.

5.3 Bodies under gravity

The notion that 'heavy' bodies tend to seek their natural resting-place at the centre of the earth enshrines the orthodox view of the philosophers of ancient Greece. The concept of 'weight', also, derives from that era, and the procedure of 'weighing' by means of the beam balance is a procedure which developed out of the scarcely rationalized intuition of the merchants and the goldsmiths of bygone civilizations. According to the older view, 'gravity', the attribute of heaviness, was a property inherent in the individual body, manifest in its tendency to fall to the earth: this tendency could only be interfered with by a 'force of resistance' preventing its realization. When the tendency was not resisted, the body fell freely with an ever-increasing 'intensity of motion'.

It is consonant with this general philosophical approach to the phenomenon of weight that the Greek natural philosophers should have overlooked, or dismissed, the possibility that there might be some simple kinematical characteristic of the free fall of bodies which is the same for all. They were not by nature experimenters – and in any case the necessary experiments are not easy to make quantitative, with simple apparatus. Weight, for them, was a property of the body, exclusively: they were naturally disposed to assert, and to believe, that the intensity of the falling motion of a body must be greater, the greater the weight of the body.

By the time of Galileo, the philosophical outlook of the Greeks was already under critical scrutiny: their geocentric cosmology had been discredited, and the notion of force, still unquantified, had largely been purified from the animistic or anthropomorphic connotations with which it originated. Ultimately, Galileo was able to write, 'The propensity of a body to fall is equal to the least resistance which suffices to support it.' In the reference to the equilibrium situation, in this statement, the weight of the body appears as having the same character as the supporting 'force of resistance' (for the two 'actions' can annul one another): in the reference to the non-equilibrium situation, consequently, the weight is seen as a force initiating the motion from rest when the body is released. Against a background of developing ideas of this kind, the question whether or not the motion of free fall 'under gravity' of different bodies is generally of the same intensity must have appeared a much more open question than it had done to the Greeks.

Galileo had, in fact, submitted this particular question, and others of a related nature, to the verdict of direct experiment during the years 1589 to 1591, when he was still a young lecturer in mathematics at the University of Pisa. By that time, as we have seen (p. 49), the concept of instantaneous velocity had been formulated – and some consideration had been given to the mathematical description of rectilinear motion with steadily increasing (or decreasing) velocity – moreover, something that we have not already mentioned, the notion of a 'perfectly smooth surface', had been adumbrated. In this last connexion, the concept of the *ideal inclined plane* had emerged as worthy of theoretical discussion, as we shall presently see. With great intuitive insight, Galileo made use of all these developments in designing his experiments.

Before we discuss Galileo's experiments, let us consider an argument, due to Simon Stevin (Stevinus of Bruges, 1548–1620), which antedates the Pisan experiments by a few years (1586). (It has indeed been claimed that the argument did not originate with Stevinus, but that it had been given by Jordanus Nemorarius some three hundred years previously, and quickly forgotten.) We imagine a solid triangular prism of principal section ABC (Figure 14) supported with its largest face AB horizontal. The base angles CAB and CBA are θ_1 and θ_2, respectively, and it is assumed that the faces AC and BC are perfectly smooth. A uniform heavy flexible band is placed around the prism and rests, as shown in the diagram, in contact with the faces AC and BC, the free portion of the band hanging below the prism. We make the intuitive assumption that in this situation the band will remain permanently at rest on the prism. To make the contrary assumption, that from an initial state of rest the band would begin to move, in one direction or the other, would involve the conclusion that the motion should continue to 'intensify' indefinitely, which would be 'absurd'.

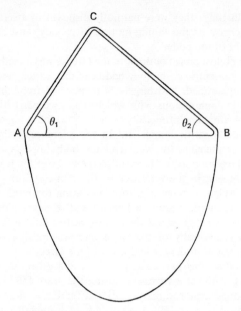

Figure 14

Adopting this assumption, then, we analyse the situation of rest. We consider separately that portion of the band in contact with the faces of the prism and the remaining portion hanging freely below it. Each portion, individually, must be in equilibrium. If the band is completely flexible, the lower portion must hang symmetrically, and for this reason the tension in the band at A must be the same as the tension at B. Also, of necessity, the tension at C must have a unique value, whether we consider its effect (in the direction $\overrightarrow{AC}$) on the portion of the band in contact with AC, or its effect (in the direction $\overrightarrow{BC}$) on the portion in contact with BC. So far as these two portions of the band are concerned, therefore, the resultant tensions, which balance the 'effective weights', W_1' and W_2', respectively, are the same, being equal to the differences of components which are equal in pairs. For a uniform band, the true weights, W_1 and W_2, of the portions in question are proportional to the lengths of the portions, that is

$$\frac{W_1}{AC} = \frac{W_2}{BC}.$$

But, from the geometry of the triangle,

$$\frac{AC}{\sin \theta_2} = \frac{BC}{\sin \theta_1},$$

thus $W_1 \sin \theta_1 = W_2 \sin \theta_2.$ **5.1**

We have already concluded that

$W_1' = \frac{1}{2} W.$ **5.2**

Obviously, equations **5.1** and **5.2** are consistent with the general result

$W' = W \sin \theta;$ **5.3**

the effective weight of a body resting on a perfectly smooth inclined plane varies with the angle of inclination of the plane according to the factor sin θ.

Throughout our discussion we have used the term 'effective weight' to refer to the action tending to produce motion down the plane (along the line of steepest descent). Equation **5.3** is not only consistent with equations **5.1** and **5.2**, but it is correctly 'normalized' in that it predicts that when the plane is vertical the effective weight is the true weight of a body. This is a self-evident proposition, for in such circumstances, if the plane is ideally smooth, the body is by definition entirely unsupported. We accept equation **5.3**, then, as a full statement of our conclusions.

We have derived our final result, as we set out to do, by giving, in its essential features, the argument of Stevinus. We have not failed to stress that this argument depends fundamentally on the initial assumption of equilibrium which at the time in question was based on nothing more secure than Stevinus's intuition. (Nowadays, we should support the assumption by reference to the principle of the conservation of energy – but 'energy' is a nineteenth-century concept, and in any case we have not as yet introduced it into our discussions.) However, it is a valid result as we now know, and out of respect for historical truth we have set it in its original historical context. To have considered it before we embark on a discussion of the experiments of Galileo, which are our main concern, will add clarity to our view of them.

It is immaterial for our purpose whether, in 1589, Galileo was aware of the formal result to which Stevinus had argued his way three years previously. What is important is that he had the same intuitive appreciation of the fact that the 'free' motion of a body on a perfectly smooth inclined plane is motion in which gravitational action is 'diluted', the more so as the inclination of the plane is decreased. This is precisely the implication of equation **5.3**, when the effective weight is regarded as the propensity of the body to 'fall' in the situation obtaining (see p. 85). Galileo wished to investigate free fall under gravity; he was acutely conscious that his means of time measurement were altogether too insensitive to make that directly possible; but his argument was that if he investigated instead the much less intense motion of bodies down inclined planes of different (small) inclinations (which he could reasonably hope to do), and if he found the same pattern of increase of distance with time whatever the inclination of the plane, then he could justifiably 'extrapolate' his experimental findings and conclude that free fall (that is, fall alongside a vertical plane) also involved the same pattern of change.

87   Bodies under Gravity

Galileo made many of his observations using a heavy plank of wood, about six metres long and twenty-five centimetres wide, in which he had cut a groove 'very straight, smooth and polished ... lined ... with parchment, also as smooth and polished as possible'. For various inclinations of this plank, he allowed 'a hard, smooth, and very round bronze ball' to roll downwards along the groove from a position of initial rest – and in successive trials he determined the time taken by the ball in covering different distances along the groove. Relative measures of elapsed time were derived from weighing the amounts of water discharged through a pipe of narrow bore soldered into the bottom of a large vessel in which the water level did not vary appreciably during the course of the experiment.

By modern standards, Galileo's arrangement was crude in the extreme, but, having a very clear appreciation of its limitations, believing that the essential features of free fall could be given simple mathematical expression, and having deduced what form that expression should take if free fall were uniformly accelerated (in the sense that we now understand this term – see p. 52), Galileo had no difficulty in deriving significant information from his measurements. He presented them as showing that, apart from inevitable small irregularities from one trial to another, the squares of his observed times were directly proportional to the distances of travel from rest, for any given inclination of the plane. This result is consistent with our equation 4.6 – and Galileo found, as he expected, that the value of the acceleration which he deduced by use of that equation increased as the inclination of the plane increased. For a fixed inclination of the plane he found, using balls of different sizes and different materials, that the acceleration was the same, within narrow limits, in all cases. From these observations, then – and extrapolating to the extreme of the vertical plane, as we have already explained – Galileo concluded that free fall under gravity is accelerated motion in which equal increments of velocity accrue in equal times, and that the measure of this acceleration is the same for all falling bodies (at a given place on the earth's surface), or, at least, that it would be so if the adventitious effect of air resistance were allowed for.

It had been the orthodox Greek view, as we have mentioned, that heavier bodies must fall more rapidly than lighter bodies in comparable circumstances. Indeed, they held that the intensity of motion should be proportional to the weight. It is interesting in this connexion to quote from the words which Galileo put into the mouth of Salviati, the spokesman for his own views, in *Dialoghi delle nuove scienze* (1638), the definitive exposition of his mechanical philosophy which was published in Leiden towards the close of his life:

Aristotle says that an iron ball of 100 pounds falling from a height of 100 cubits reaches the ground before a one-pound ball has fallen [more than] a single cubit. I say that they arrive at the same time. You find, on making the experiment, that the larger outstrips the smaller by two finger-breadths ..., now you would not hide behind these two fingers the ninety-nine cubits of Aristotle. ...

We need not labour the point, for our claim will assert itself time and again as our discussion proceeds, but it must be clear from this instance alone that not the least of the aspects of Galileo's genius was that, as no one before him had done, he recognized the impossibility in the real world of excluding every extraneous circumstance and working with the type of completely isolated system which of necessity we must assume in our efforts to theorize. This realization, as we have already stated (p. 13), is absolutely fundamental if we are to develop a satisfactory philosophy of our science.

Before we consider Galileo's other experiments with the inclined plane, there is one matter concerning his 'extrapolation to the vertical' to which we should refer. His experiments had been performed with spherical bodies rolling down planes having various small inclinations to the horizontal. Strictly, extrapolation to the vertical results in an ideal situation in which a ball 'rolls down' the surface of a vertical plane. In such motion without slipping the ball at any stage in its motion is rotating about a horizontal axis with angular velocity v/r, where v is the instantaneous linear velocity of its centre in the downwards direction and r is its radius. In free fall under gravity, on the other hand, there is no such (accelerated) rotation. Galileo did not resolve this 'objection'; indeed it was only with the development of Newtonian mechanics that it could be resolved – and shown to be of no consequence.

Galileo's first experiments, as we have seen, established the 'acceleration due to gravity' as a characteristic quantity (for any point on the earth's surface), the same for all bodies whatever their weight. His second series of experiments established – or almost established – the 'principle of inertia' as we know it today. Essentially, in this series, Galileo placed two inclined planes in opposition, forming a shallow V, so that there was, in effect, a continuous groove, down the one plane and up the other. His intention was to see how far a ball, released from rest on the first plane, would roll up the other before coming instantaneously to rest on that plane. Keeping the inclination of the first plane constant, he varied the inclination of the other plane, as also the initial point of release of the ball. Inevitably, there was a measure of irregularity in his observations, as before, but he convinced himself of a simple result. He concluded that, in the ideal situation, with perfectly smooth surfaces, and in the absence of air resistance, the points of release on the first plane and instantaneous rest on the second plane would in every case be at precisely the same height above any horizontal reference plane set up for purposes of comparison.

By definition, an inclined plane is a surface of constant 'steepest slope'. The result that we have just described has reference specifically to the particular arrangement of two such surfaces in opposition. Galileo proceeded to generalize this result, intuitively, to the case of surfaces of variable steepest slope. Suppose that we have a perfectly smooth cylindrical surface so disposed that the generating lines of the surface are horizontal. Galileo asserted that a ball released from any point on such a surface (provided that it remained in contact with the surface throughout its motion) would roll down the surface, cross the lowest generating line perpendicularly, and continue forwards and upwards until it came momentarily to rest at the same height 'above ground' as that from which it started.

In this connexion, also, Galileo saw the motion of the spherical bob of a simple pendulum in plane oscillation as dynamically equivalent to the 'free' motion of a sphere in a circular cylinder (the axis of the cylinder being horizontal and its radius being equal to the length of the pendulum). Although the physical constraints were different in the two cases, the path of the moving body was essentially the same in each; according to Galileo's intuition the character of the motion had to be the same also. In any event, it is a fact of observation that the oscillations of a freely swinging pendulum are very closely symmetrical about its mean position: in the ideal situation, in the absence of air resistance, and with a perfectly flexible suspension, according to Galileo a simple pendulum would continue to oscillate with constant amplitude indefinitely – the bob would rise to the same height at each extremity of its swing in either direction.

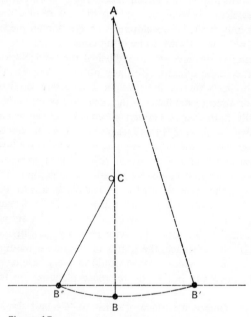

Figure 15

Pursuing this line of thought, Galileo devised the following experiment. He suspended a heavy bob (B, Figure 15) by means of a long flexible string from a rigid support at A. Hanging vertically, the string passed alongside a securely fixed peg at C. Having drawn the bob aside, with the string taut, to the position B' (lower than C), Galileo observed the motion of the bob once it was released. As the bob passed through the lowest point of its swing, the motion of the string was interrupted by the peg, but the bob continued to B″, the string remaining taut throughout. As Galileo had predicted, the line joining B″ and B′ was horizontal, nearly enough, in all repetitions of the experiment. He would have obtained a similar result, so he surmised, if, instead, he had used a composite cylindrical surface, having different (cylindrical) radii over the portions B′B and BB″, and, dispensing with the constraint of the suspension, he had allowed a ball to roll from rest at B′ down the line of steepest descent. We need only note, in accepting this intuitive conclusion as valid, that Galileo was ignoring the complication of rotation as he had done in his argument regarding free fall in relation to his first experiment (p. 89). Again, however, the soundness of his conclusion was unaffected by this neglect.

We have said that Galileo's second series of experiments established – or almost established – the modern 'principle of inertia'. His argument was straightforward. The 'effective weight' of a body resting on an ideally smooth horizontal surface is zero (equation 5.3) – in these circumstances the body is free from the action of any horizontally directed force, due to contact with the plane or to its own weight. Suppose, then, that a body rolls down an ideal inclined plane on to such a horizontal surface: according to the experiments it must continue to move until it reaches a point at the same height as its original point of release. In fact, it must continue indefinitely, since this endpoint is for ever unattainable. Generally, therefore, a body entirely free from the action of any (resultant) force must continue to move with constant speed indefinitely (and, if originally at rest, it must remain at rest).

We use the word *inertia* to designate this universal attribute of bodies, first securely identified in Galileo's conclusion, that they do not change the character of their motions in the absence of force. This conclusion is in direct opposition to the Greek view that force is necessary to maintain motion. In the entire absence of force, according to the view of the Greek philosophers, all motion, save perhaps the 'perfect' motion of the stars in the firmament, must ultimately cease.

Galileo's principle of inertia, as it was finally formulated, was indeed in direct opposition to the Greek view in relation to terrestrial motions (from the observation of which it was in fact derived), but, strangely enough, it did not break completely with the older view regarding extra-terrestrial motion. The Copernican cosmology, which Galileo had accepted, had shifted the centre of reference from the earth to the sun, but it had not abandoned the notion of uniform motion in a circle as perfect motion in respect of the heavenly bodies generally (p. 40). Neither did Galileo abandon this notion: he adopted it gratuitously in his attempt to generalize the conclusion which he had reached in relation to his laboratory experiments – experiments on 'local motion' as he called them – and so formulated his principle of inertia in respect of large-scale motion on the earth's surface that in the end he came to assert what we now know to be untrue. He imagined the perfectly smooth horizontal surface of limited extent, which had served as an ideal component in the 'thought experiment' of the laboratory, to be extended until it circled the earth along a great circle (p. 74). Extrapolating from his essentially valid conclusion in relation to local motion, he asserted that along such a great-circular path, a body of whatever initial velocity would remain in contact with the surface, travelling around the path with constant speed, indefinitely. In this case Galileo's intuition misled him – and by reason of this lapse his earthbound principle of inertia failed, in the end, to achieve full generality.

The first philosopher explicitly to condemn the ancient distinction between terrestrial and extra-terrestrial motion was René Descartes (p. 16). He sought to build a model of the whole universe on the basis of a single set of ideas. He expounded these ideas in *The Principles of Philosophy* published in Amsterdam in 1644, six years after Galileo's *Dialogues on the New Sciences* had appeared in Leiden. It must be stressed at the outset that Descartes' system was a speculative system, owing nothing to any experiments of his own, for he made none – and little to those of others, but it had a powerful influence on contemporary thought and attracted many adherents. For our purposes we only need to remark that it involved a principle of inertia different from Galileo's – one that was accepted by Newton, and is effectively accepted today. We may indicate its nature by contrasting, in quasi-philosophical terms, Descartes' view with that of the Greeks. If we understand by 'natural' or 'perfect' motion that type of motion which does not require explanation in terms of a contingent cause, then, in respect of an earth-centred cosmos, the Greeks regarded uniform motion in a circle as perfect, whatever the speed – Descartes, in respect of a material universe existing in infinite Euclidean space, regarded uniform motion in a straight line as perfect, at any speed.

5.4 Bodies in collision

Galileo's investigations, as we have described them in the last section, dealt exclusively with the changes in the motion of single bodies brought about by the action of their own weight – that is, by the 'force of gravity' acting on them – this force being 'diluted' by the support due to an inclined plane, or a pendulum suspension. Once the essential features of this experimental situation had been elucidated, it was a natural development to introduce a second experimental body and to investigate the mutual changes of motion of two such bodies resulting from their collision. Thus, a 'second' ball might be placed at rest at the lowest point of the shallow-V groove of Galileo's opposed inclined planes, and the effect of allowing the 'first' ball to roll, from a given point of release, down one of the planes, so as to strike the other ball, might be examined. Alternatively, two balls might be suspended by parallel threads of equal length so as to be in contact when the suspensions were vertical, and one ball might be displaced sideways and released, so providing a similar collision situation. Although the more accurate investigations of motion under (diluted) gravity had been made by Galileo with inclined planes rather than with pendulum suspensions, as we have already described, it is a fact of history that the early experiments on collisions, in the period from 1660 to 1680, were generally made using the two-pendulum arrangement. In these experiments, with collisions occurring only under conditions in which both balls were instantaneously at the lowest points of their respective arcs of swing, it was obvious (see p. 91) that gravitational forces had no influence on the 'sudden' changes of velocity taking place at the moment of collision. In modern phraseology we should say that the changes were due to 'elastic' forces brought into play during the (extremely short) time for which physical contact between the colliding balls was maintained.

In respect of the ultimate development of the science of dynamics, the importance of the collision experiments was simply that they provided for the first time direct quantitative information regarding the effects of forces other than gravitational forces – and more significantly so because the forces concerned were 'mutual' forces developed within a laboratory system (of two bodies) which could reasonably be regarded (during the collision phase, at least) as an isolated system.

From the practical aspect, the two-pendulum arrangement used in these experiments had the great convenience that velocities before and after impact could be evaluated in relative measure directly by measurement of distance, without recourse to the more difficult measurement of time. This possibility derives from the previous experiments of Galileo. Galileo's experiments had shown, in effect, that the speed acquired in motion down an ideal inclined plane, or along the arc of a circle when the moving body is supported by an ideal suspension, depends only on the vertical distance traversed by the body from rest; it is the same as the speed acquired in free fall under gravity through the same distance. This speed (for uniformly accelerated motion) is proportional to the square root of the distance of fall (equation 4.7). In the pendulum arrangement, moreover, it is a simple geometrical result that, for angular displacements of no more than a small fraction of a radian, the square root of the vertical component of linear displacement of the suspended body is directly proportional to the first power of the horizontal component of displacement. It follows that, in this arrangement, the speed of the body at the lowest point of its path (for small angular displacements) is directly proportional to the horizontal distance through which the body has moved since release (or through which it will move before again coming instantaneously to rest). Measures of horizontal displacements, in a two-pendulum arrangement, serve, therefore, as relative measures of the velocities of the suspended bodies at the point of collision, provided always that this occurs when the bodies are moving horizontally.

During the period from 1660 to 1670 extensive experiments on collisions were made, using the two-pendulum arrangement, by Edme Mariotte (c. 1620–1684), in France, and by Christopher Wren (1632–1723), in England. The problem also engaged the attention of Christiaan Huygens (see p. 37) and John Wallis (p. 21). Wallis devoted a section of his *Mechanics* (1669–71) to its consideration, and was the first clearly to enunciate the modern view that the forces brought into play at the moment of collision are related to the instantaneous deformations of the colliding bodies in the region of contact. Following the investigations of Wren and Wallis (Mariotte's account of his experiments was not published until after his death), Newton carried out further experiments of a similar character at various times during the succeeding years. Soon afterwards (December 1684) he had given a first sketch of a new system of dynamics in a small treatise *De Motu Corporum*, and by 28 April 1686 the manuscript of the first book of *Principia* was ready for the press. By that time the essential concepts of mass and force had been given their final form, and the experimental results on collision phenomena had been assimilated into the over-all picture, finding their natural place in the presentation, without any special emphasis or distinction.

In the introduction to this chapter (p. 83) we asserted that the Newtonian concepts of mass and force would not have been developed in the way that they were developed if there had 'not already been on record the experimental observations' on collisions and falling bodies. We have to distinguish between the two components of this assertion. There is no difficulty about the observations on falling bodies. Galileo's observations are adequately documented, and a principle of inertia had been built out of them, before the time of Newton, as we described in the last section. The observations on collisions, on the other hand, were made during Newton's lifetime, the more significant of them by Newton himself, when he already had the beginnings of the mass concept in his mind. There is strong circumstantial evidence that his final formulation of this concept was coloured by the results of these collision experiments, as we shall seek to show in the next chapter, but there is no simple conclusion of wide generality, similar to the principle of inertia, which he admitted having derived directly from them. In spite of this difficulty, we shall attempt to reconstruct some of the experimental results which were available to Newton in this domain – and the empirical regularities which they exhibited – for, in whatever way the concept of mass actually matured in his thinking, for our purposes it may best be approached by this route. We make no attempt to give a literal description of original experiments, or a strict assignment of priorities as between the various experimenters (for not all the information is available): our description must be regarded as expository rather than historically valid.

Suppose, then, that we consider the collision of two spherical bodies, A and B, confining our attention to the case in which B is initially at rest and the entire motion (of the centres of the bodies), both before and after collision, is in a single horizontal straight line. We take the direction of the initial velocity of A as the positive direction along this line – and we assume that if v_1, the magnitude of this velocity, is given, then the magnitudes of the final velocities of A (v_1') and B (v_2') must be uniquely determined. Essentially, the kinematical problem involves nothing more than the elucidation of the relationships between v_1' and v_1, and v_2' and v_1, for the arrangement proposed. Regarded in this light, the experiments showed that, for all pairs of bodies, the empirical relationships were of direct proportionality, exclusively. Expressing these results in symbols, we have $v_1' = k_1 v_1$, $v_2' = k_2 v_1$, k_1 and k_2 being constants, when v_1 is varied, for a given pair of bodies – and we note that k_2 always appeared as a positive constant, and k_1 positive in some cases and negative in others, but always in the range between $+1$ and -1, when different pairs of bodies were investigated.

On the basis of this simple result, which at first sight might appear to have been inevitable, many other constant ratios involving the three magnitudes v_1, v_1' and v_2 may be constructed. Newton identified two of these ratios as particularly significant; the first was the ratio $(v_2' - v_1')/v_1$, the second the ratio $v_2'/(v_1 - v_1')$. The former ratio is the ratio of the relative velocity of the two bodies after collision (B leading) to the relative velocity before collision (A 'leading'), the latter is the ratio of the increase in velocity of B to the decrease in velocity of A. In terms of k_1 and k_2, these ratios are given as $k_2 - k_1$ and $k_2/(1 - k_1)$, respectively. The second ratio is necessarily a positive quantity for all pairs of bodies (because both k_2 and $1 - k_1$ are positive in all cases); the first, $k_2 - k_1$, was shown by experiment likewise to be positive and never to be greater than unity.

Newton referred to the first ratio $(v_2' - v_1')/v_1$, as the *coefficient of restitution* effective in the collision: it represents the fraction of the (relative) velocity of approach which is restored in the (relative) velocity of separation of the bodies after the collision. As we have said, for a given pair of bodies (at least, for not-too-violent collisions) this quantity is constant for all values of the relative velocity of approach. Further investigation showed that the coefficient of restitution is characteristic of the pair of materials of which the bodies are made – the materials being given, the value of the ratio is the same independently of the sizes of the bodies. Taking up the notion which Wallis introduced (see p. 93), we conclude that the value of the coefficient of restitution is some (possibly not very simply specified) index of the type of deformation–force relationship characterizing the physical processes involved at impact, when the materials in question are given. (Throughout this paragraph, speaking of 'materials', in the plural, we do not, of course, exclude the possibility that the colliding bodies may be made of the same material.)

Newton's second ratio $v_2/(v_1 - v_1')$ turned out to be of much greater interest and significance than his first. When the bodies A and B were made of the same material, and were solid throughout, this ratio proved to be equal to the ratio of the volume of A to the volume of B, within experimental uncertainty. Formally, if V_1, V_2, represent the volumes, in this case

$$\frac{v_2'}{v_1 - v_1'} = \frac{V_1}{V_2}.$$

Only a single material being involved, and the bodies being homogeneous, the ratio of the volumes is also the ratio of the quantities of matter in the two bodies (ultimately, the ratio of the numbers of atoms in each). In this case, therefore, B gains velocity and A loses velocity, and B's gain is to A's loss as the quantity of matter in A is to the quantity of matter in B.

Now, it is axiomatic that the beam-balance of equal arms provides a means of judging the equality of weight of two bodies (the equality of the forces required to support them – see p. 85). On the basis of experiments with such a balance, using homogeneous solid bodies all made of the same material, we may demonstrate that the weight of a body is directly proportional to its volume (thus instances might be multiplied showing the equality of weight of n bodies, each of volume V, with a single body of volume nV).

When the collision experiments were carried out with two bodies made of different materials, it turned out, again within experimental uncertainty, that the ratio $v_2'/(v_1 - v_1')$ was in every case equal to the ratio of the weight of the body A to the weight of the body B, as determined by the balance.

There are two ways of summarizing all these results concerning the ratio $v_2'/(v_1 - v_1')$. We might merely assert that in all cases, whether the two bodies are made of the same or of different materials, the gain of velocity by B is to the loss of velocity by A as the weight of A is to the weight of B. However, this assertion would not increase our understanding of the phenomenon, for the action of gravity is excluded in the collision situation, as we have already agreed (p. 93) – and weight is essentially an attribute of gravitational action. Or, alternatively – and this is the other way – basing our evaluation more particularly on the results obtained with bodies made of the same material, we might make the bolder assertion that in all cases, whatever the materials of which the bodies are made, the ratio $v_2'/(v_1 - v_1')$ is equal to the ratio of the quantity of matter in A to the quantity of matter in B. If we make this assertion (as, in effect, Newton eventually made the assertion in Book 1 of *Principia*), then we have introduced a new measurable quantity into physics: 'the quantity of matter in a body of whatever chemical composition' – a quantity different in nature from the weight of the body (but in some way intimately involved in the weight as determined by the laboratory balance).

At this stage we are already anticipating the developments of the next chapter, so let us conclude this section by assigning the name *inertial mass* to the quantity of matter defined as we have just defined it in relation to the particular collision situation that we have been discussing. If we do this, and adopt the symbol m to represent this new quantity, the empirical result on which we have based this definition may be written

$$\frac{v_2'}{v_1 - v_1'} = \frac{m_1}{m_2},$$

or $\quad m_2 v_2' = m_1(v_1 - v_1').$ **5.4**

We note that equation **5.4** has simple formal significance. The left-hand member of this equation specifies the increase for body B, and the right-hand member the decrease for body A, as a result of the collision, of the composite quantity represented by the symbol mv. Newton was to define this composite (derived) quantity as the *quantity of motion* of a body (see p. 102). In modern phraseology it is referred to as the *linear momentum*. Equation **5.4** states that, in the direct collision of two bodies, one being initially at rest, the gain of momentum by the struck body is precisely equal to the loss of momentum by the striking body. We postpone the generalization of this result for later consideration (p. 136).

Further reading

H. Butterfield, *The Origins of Modern Science, 1300–1800*, (Chapter 5), Bell, 1949.

H. Margenau, *The Nature of Physical Reality*, McGraw-Hill, 1950.

A. N. Whitehead, *Science and the Modern World*, Cambridge University Press, 1926.

W. P. D. Wightman, *The Growth of Scientific Ideas* (Part I), Oliver & Boyd, 1951.

Chapter 6
Mass and Force

6.1 Towards a quantitative definition of force

The word 'force' was widely used in the writings of natural philosophers before the time of Newton, even before the time of Galileo. However, at that stage, the force concept was essentially unquantified: no all-embracing definition had been devised in terms of which a measure could be assigned to the force exerted in any specified set of circumstances. The concept had had its origin in common experience, in recognition of the effort which a man must make if he wishes to move, or to arrest the movement of, any massive body with which he has to deal. In such a human situation, we are conscious of a tensing of the muscles; in the ordinary way of speaking we say that we have to exert force to achieve our aims.

The principle of inertia – whether accepted in its Galilean or its Cartesian form (see p. 92) – provided the first clarification of this confused situation: it described the ideal behaviour of a body totally 'shielded' from the action of force of any kind. Under these conditions (which are strictly unrealizable in practice), the principle asserts that the body maintains its existing state of motion without change. Once this principle had been accepted as basic for the science of dynamics, the most urgent problem became that of devising a definition of force in terms of the change of motion produced.

Galileo showed that the effect of the weight of a body, in relation to free fall, was continuously and uniformly to accelerate the velocity of fall of the body. This, he asserted, was uniquely the effect of the force of gravity acting on the body continuously. Bodies differed, one from another, in weight, that is in the magnitude of the force of gravity acting on them, but, under ideal conditions, the acceleration was the same in all cases. Obviously, in this situation, the measure of the constant acceleration is the simplest purely kinematical measure of the (continuous) change of motion of the falling body. Galileo was predisposed to the view that the mathematical description of natural phenomena would eventually prove to be possible in the simplest terms, but he failed to devise a general definition of force in terms of acceleration, nevertheless. We may conjecture, possessing the advantage of hindsight, that his difficulty was that for him the attribute of inertia was still unquantified: his inertial principle was no more than a qualitative statement, in spite of its fundamental importance.

Newton, as we have said, had the experiments on impact, as well as those on falling bodies, out of the results of which to fashion his concepts. Moreover, he had definitely adopted the principle of inertia in its Cartesian form (p. 92). In the collision experiments, changes of motion of each of the colliding bodies have to be considered. These changes are essentially 'sudden', and such forces as act are effective only during the very short duration of physical contact between the bodies that collide. The intimate details of the collision process were not open to Newton's investigation – only the end results. In this situation the over-all change of velocity of each body was the obvious kinematic quantity to determine (in section 5.4 we described the experimental results on this basis). Such a change of velocity may be regarded as resulting from the effect of a time-varying acceleration (having a non-zero value only during the duration of contact). In the notation of section 5.4, for example,

$$v_1' = v_1 + \int_0^\tau a_1 \, dt, \qquad\qquad 6.1$$

where a_1 is the acceleration of body A, and τ is the duration of contact in the collision. According to Wallis (1671), each body exerts force on the other, during the time τ: these forces increase as the contact deformation increases, and decrease to zero again when the collision is complete. In modern phraseology we have referred to these forces as 'elastic' forces (p. 93); Wallis described them as due to the 'spring' of the materials of which the bodies were made. We note in passing that about this time Robert Hooke (1635–1703), having had ample experience of the properties of 'real' springs as components of mechanism, enunciated the principle (1676) that deformation in such cases is in general proportional to the force applied. Whilst Hooke's experience had obvious relevance to the theoretical notions of Wallis, we must be clear that, at least at the time of its formulation, 'Hooke's law' was entirely extraneous to the subject of dynamics. It might possibly form the basis for a statical definition of a so-called 'force', but it did not provide, or even contribute to, a definition of force as an agent responsible for accelerated motion.

When Newton came to formulate the basic laws of dynamics (see section 6.2), it eventually became clear that he had introduced an altogether bold and significant addition to Wallis's views of the collision process. He had asserted, on intuitive rather than on logico-deductive grounds, that whatever the nature of the deformations of the two bodies – whether they are equally deformed, or one only slightly and the other considerably deformed in the collision, the force which each exerts on the other across the surface of contact at any instant is the same. So the time integrals of these forces over the period of collision are the same:

$$\int_0^\tau {}_1F_2 \, dt = -\int_0^\tau {}_2F_1 \, dt. \qquad\qquad 6.2$$

In equation 6.2 ${}_1F_2$ represents the force, in the direction of the line of centres, acting from body A on body B, and $-{}_2F_1$ the (equal) force acting in the opposite direction from body B on body A. In making this assertion Newton had not immediately provided a formal basis for the quantification of the force concept: he had merely put forward a physical hypothesis (in general form, see p. 103), which subsequent experimental investigation was to elevate into a 'law of nature'. With the aid of this law he was able to devise quantitative definitions of both force and inertial mass, so making it possible (in relation to our present restricted field of inquiry) to bring the results of the experiments on falling bodies and those on collisions into the same scheme of interpretation. We may mention here that in modern phraseology we refer to the time integral of the force, as represented by either member of equation 6.2, as the *impulse* generated in the collision.

In order to elucidate the statements that we have just made let us consider further equations 5.4, 6.1 and 6.2. We first supplement equation 6.1 by the corresponding equation for body B (we are using the notation of section 5.4 throughout). We then have

$$v_1 - v_1' = -\int_0^\tau a_1 \, dt,$$

$$v_2' = \int_0^\tau a_2 \, dt.$$

From these two equations we obtain an expression for the ratio $v_2'/(v_1 - v_1')$, and, because of the equality represented by equation 6.2, we may write it in the form

$$\frac{v_2'}{v_1 - v_1'} = \frac{\int_0^\tau {}_2F_1 \, dt \bigg/ \int_0^\tau a_1 \, dt}{\int_0^\tau {}_1F_2 \, dt \bigg/ \int_0^\tau a_2 \, dt}. \qquad \qquad 6.3$$

According to experiment this ratio is constant for a given pair of bodies. Anticipating the discussion of this section, we have already (p. 96) designated the ratio as the inverse ratio of the 'inertial masses' of the bodies concerned. Equation 5.4 symbolizes that interpretation. Taking equation 5.4 together with equation 6.3 we now have the formal result

$$\frac{\int_0^\tau {}_2F_1 \, dt}{\int_0^\tau {}_1F_2 \, dt} = \frac{m_1 \int_0^\tau a_1 \, dt}{m_2 \int_0^\tau a_2 \, dt}. \qquad \qquad 6.4$$

If, in general, for any body in collision with another, $F = ma$, in terms of the definitions that we have given, then equation 6.4 is obviously satisfied: in the context of the physical situation that we are discussing it is difficult to see how it could be satisfied on any other assumption.

We recapitulate at this stage. Accepting Newton's intuitive (and quantitative) extension of Wallis's views regarding the forces developed in collisions, and his (equally intuitive) introduction of a quantified mass concept in this context, on the basis of the not entirely logical chain of argument which we set out in section 5.4, we have been led to accept the general result that, in respect of the collision process, at least, these two postulates can be assimilated with one another if and only if force is measured by the product of the measures of the inertial mass and the acceleration concerned. We have still to show that these same postulates may be assimilated, on the same basis, in the experimental context of free fall under gravity. This we now proceed to do.

According to experiment, as we have noted, bodies of whatever inertial mass, in free fall, experience the same acceleration 'due to gravity'; also, if one body is in collision with another at rest, the ratio $v_2'/(v_1 - v_1')$ is numerically equal to the inverse ratio of the weights of the bodies.

The latter experimental result is represented by the equation

$$\frac{v_2'}{v_1 - v_1'} = \frac{W_1}{W_2};$$

when we bring in Newton's first postulate (represented by equation 5.4), we have, instead,

$$\frac{W_1}{W_2} = \frac{m_1}{m_2}. \qquad\qquad 6.5$$

The former experimental result, on the basis of the collision-based definition of force which involves both postulates, justifies the general symbolic statement

$$W = mg \qquad\qquad 6.6$$

(here g is the constant acceleration due to gravity, and W, the weight of the body, is the force giving rise to this acceleration in a body of inertial mass m). Clearly, equations 6.5 and 6.6 are mutually consistent: we have established the point that we set out to make – the two postulates which Newton introduced make it possible to begin to understand the two phenomena, of free fall and two-body collisions, on the basis of definitions of force and inertial mass which together lead to the quantifying equation $F = ma$.

We have made it clear, throughout the discussions of this section, and at the end of the last chapter, that we have been approaching the problem of the formulation of the fundamental concepts of dynamics by considering in depth two particularly simple laboratory experiments. Our considerations will have demonstrated that even in this well-defined situation there was no high-road pointing the way of advance to the first competent investigator who might give serious attention to the problems involved. It is probable that much of the ground that we have traversed in these discussions was traversed by Newton during the decade or so before the writing of *Principia*. However, such considerations formed but one strand of his thinking: he was concerned as much with the planets in their orbits as with stones falling to the ground, or pendulums oscillating in his laboratory. When he came to set out his conclusions in writing he did so almost defiantly – admitting no situation, on the earth or among the stars, to which they did not apply: in setting them out he suppressed every trace of the devious routes by which they had been attained, and every hint (save possibly one – see p. 106) of the extraordinary intuition which informed them.

Having admitted the inadequacy of disciplined logic – or of 'the scientific method', however that be defined – for the eliciting of significant conceptual novelty from the simple laboratory results which we have been considering, we shall certainly not attempt to continue the same procedure in a wider context: in the next section, therefore, without further ado, we shall simply give Newton's formulation of the general 'laws', and reserve our further comments until we can see his whole majestic edifice in one piece, as it was presented in the Introduction to Book 1 of *Principia* in 1686.

6.2 Newton's ' laws of motion '

Newton's *Philosophiae Naturalis Principia Mathematica* was written in Latin. Perforce, therefore, we give here a rendering in English. For this purpose we adopt the translation of J.H. Evans and P.T. Main: it was made in the mid-nineteenth century, at a time when an edited version of the more elementary portions of Newton's great work was in use as a standard textbook for university students of physics in Britain.

The statements to which Newton gave the name 'Axiomata, sive Leges Motus' are three in number. These are the three 'laws of motion' which are paraded in modern form in the elementary textbooks of today. In any serious presentation, however, these three statements should not be divorced from the eight statements which preceded them in the original version: the eight prefatory statements were referred to as 'Definitiones' by Newton. We give here, therefore, Newton's eight definitions, then his three laws (or axioms).

Definition 1. Quantity of matter is the measure of it arising from its density and bulk conjointly.

Definition 2. The quantity of motion of a body is the measure of it arising from its velocity and the quantity of matter conjointly.

Definition 3. The innate force of matter is its power of resisting, whereby every body, so far as depends on itself, perseveres in its state, either of rest, or of uniform motion in a straight line.

Definition 4. An impressed force is an action exerted on a body, tending to change its state either of rest or of uniform motion in a straight line.

Definition 5. A centripetal force is one by which bodies are drawn, impelled, or in any other way tend from all parts towards some point as centre.

Definition 6. The absolute magnitude of a centripetal force is a measure of it which is greater or less according to the efficacy of the cause which propagates it from the centre through the regions of space all round it.

Definition 7. The accelerative quantity of a centripetal force is a measure of it proportional to the velocity which it generates in a given time.

Definition 8. The motive quantity of a centripetal force is a measure of it proportional to the motion which it generates in a given time.

Newton added short explanatory or illustrative passages after each definition; we reserve comment to the end, merely noting here two aspects of modern usage. In Definition 2 we should now write, instead of 'quantity of motion', 'momentum' (see p. 97) – and in Definition 3, instead of 'innate force', 'inertia'. We proceed, then, to quote the 'axioms or laws'.

Law 1. Every body perseveres in its state of rest, or of uniform motion in a straight line, except in so far as it is compelled to change that state by forces impressed on it.

Law 2. Change of motion is proportional to the moving force impressed, and takes place in the straight line in which that force is impressed.

Law 3. An action is always opposed by an equal reaction; or, the mutual actions of two bodies are always equal and act in opposite directions.

In the same introductory section of *Principia* in which the eight definitions and three laws were set out, Newton also gave six corollaries. In whatever sense we may understand the categories of 'definition' and 'law' (see p. 104), that of 'corollary' is surely unambiguous. It implies a statement which is deducible by strictly logical argument from other statements previously accepted as valid. If this were true of each of Newton's corollaries, there would be no need to quote them here. It might be profitable to refer to some of them at a later stage of the discussion, but, if we were to quote them at the outset, they would have nothing to contribute to the fundamental structure of the system that we are about to discuss. As it happens, the situation is otherwise. Critical assessment of the position shows that Newton's first corollary is indeed essential to his whole theoretical picture: we therefore quote it at once, before proceeding to our discussion.

Corollary 1. By the combined action of two forces a body will describe the diagonal of a parallelogram in the same time as the sides would be described by the body under the action of each force separately.

6.3   Criticism of Newton's formulation

It is a perpetual problem for the scientist, to develop a new language whenever his science enters on a new phase of expansion – extending its horizons to bring new aspects of natural phenomena under quantitative scrutiny. Necessarily, such a language, to be effective, must involve a considerable component of mathematical symbolism rendered significant by the formal equations which enshrine the quantitative definitions of relevant concepts – and, here and there (see p. 14), a 'law of nature' (which might have been otherwise in a conceivable universe, but in fact is unique in the universe as we know it). These formal equations constitute and exhibit the grammar of the language which the scientist develops.

In ordinary life, languages evolve slowly and are modified by use; later, they are subjected to the analysis of scholars and the underlying grammar becomes manifest. It cannot be altogether different with the languages of science, though they develop more rapidly. Some familiarity in use, an evolutionary period of trial and error, is necessary before the grammar can be established. In relation to dynamics, through the long pre-Newtonian era, the first phase of this evolution of language was taking place. Newton's achievement was effectively to telescope the second and more significant phase of its evolution by use, and the writing of its grammar, into little more than a decade of years. It was a monumental achievement, definitive in a practical sense for more than a couple of centuries and still largely significant. Inevitably, however, in the eyes of the grammarians of later generations, it was not immune from criticism. A man talking almost exclusively to himself, as Newton was, could hardly be expected to produce a universally comprehensible language – and the structure of his grammar was not fully articulated. In this section we join the ranks of the latter-day grammarians, and attempt to explore the criticisms of Newton's formulation which we have indicated in the last sentence.

Let us concede, at the outset, that the Newtonian scheme was highly successful: on the macroscopic level – once it was understood – it appeared to reflect the kinetic behaviour of the real world with complete fidelity. In later developments, on the molecular level, it proved to be an adequate basis for the initial formulation of the kinetic theory of gases. What is in question is not the success of the scheme, but its structure: we wish to decide how much of the structure is simple definition and how much is a statement of natural law. Obviously, much is mere definition; equally obviously, there must be a significant component expressing truths about the real world, for the scheme matches with reality. If possible, we wish to separate these components of definition and law – thereby increasing our understanding of the world in which we live.

Looking at Newton's formulation from this point of view, our first conclusion is self-evident: Newton's own categorization of his eleven formal statements as 'definitions' and 'laws' is grossly misleading. We may point to one obvious example of 'distinction without apparent difference', to illustrate this conclusion, before we proceed to discuss it systematically.

Newton's Definition 4 purports to give precision to the technical term 'impressed force' which is subsequently to be used (in a verbal rather than an adjectival variant) in Law 1. However, if we substitute the words of the definition into the law, we have nothing more than the seemingly tautological statement, 'Every body perseveres in its state of rest, or of uniform motion in a straight line, except in so far as it is compelled to change that state by actions which, when exerted on a body, tend to change its state either of rest or of uniform motion in a straight line.' Perhaps in this form the statement is not completely tautological, but if it has any significant content at all, surely that content is more tersely expressed by Definition 3, which we give here in a more modern version, avoiding the phrase 'the innate force of matter' which we have already noted as archaic (p. 103). In its revised version, Definition 3 reads, 'All matter has the quality of inertia, whereby every body, so far as depends upon itself, perseveres in its state, either of rest or of uniform motion in a straight line.'

On any view, then, only two of the three statements in question are independent statements – and there must be considerable doubt whether any part of the over-all content of these statements qualifies under the category of 'law'. The most that we can say in defence of Newton in this connexion is that, in writing the introduction to *Principia*, he was still practising the use of his new language, for the purpose of familiarizing his readers with its vocabulary, leaving the grammar to become manifest in the process.

We proceed now to the more systematic discussion of our stated conclusion. For this purpose we confine attention to those of Newton's statements which appear to provide recipes giving the measure of one physical quantity in terms of the measures of other such quantities. Qualitative statements are entirely admissible in the definition of new concepts, but until recipes are provided whereby values may be assigned to the measures of physical quantities representative of those concepts little of significance has been added to our possibilities of understanding. Physics is essentially the science of measurement and calculation.

The quantifying recipes appear to be Definitions 1 and 2 (and, in a less precise sense, Definitions 7 and 8) and (again lacking full precision) Law 2. Definition 1 has the appearance of providing a recipe for calculating the measure of the mass of a body. (Newton's gloss on this definition contains the statement, 'This quantity of matter is, in what follows, sometimes called the body, or mass.') Definition 2 has the appearance of providing a similar recipe for the measure of the momentum (we have already noted this modern synonym for 'quantity of motion'). Let us consider these two definitions before we proceed farther.

It is immediately clear that if mass has been satisfactorily defined by Definition 1, then Definition 2 is a perfectly satisfactory definition of momentum (we leave aside the question of the vector or scalar characteristics of the new quantities for the moment). With mass already defined, Definition 2 is a typical example of the quantifying definition of a new 'derived' quantity (see p. 32) in terms of other quantities which are experimentally determinable. The representative equation ($p = mv$, in this case, in the standard notation) is such that in any given situation, on the basis of measurements made, values may be assigned to all but one of the symbols involved in the equality, so that the value attached to the remaining symbol (representing the measure of the new quantity) may be calculated.

On the other hand, it is equally clear that Definition 1 (unsupported by any previous definition) entirely fails to satisfy the same requirements as a definition of mass. In the standard notation, the representative equation in this case is $m = \rho V$. We assume that V, the volume (bulk) of the body, may be determined experimentally, but no procedure has been laid down for the measurement of density ρ. Indeed, we are naturally doubtful whether mass can be defined in any way as a derived quantity, when the only 'primary' quantities that we have hitherto recognized are those of length and time. In taking up the science of dynamics we are consciously entering the domain in which the intrinsic properties of material bodies are involved – and 'real' phenomena; space and time (at least in the Newtonian view) merely provide the stage on which the action occurs. We naturally assume that at least one new primary quantity has to be identified before we can develop the quantitative side of this science successfully. On this basis we reject Definition 1 (though it is a 'true' statement, defining density once mass has been defined) – or any other similar definition – as a definition of mass.

In his comments on Definition 1, Newton wrote, '[The mass] is known for each body by means of its weight; for it has been found, by very accurate experiments with pendulums, to be proportional to the weight.' Though he gave no particular prominence to this statement, in fairness to Newton, we ought to examine the possibility that a satisfactory definition of mass might be based upon it. Leaving aside the necessity for a definition of density (surely a trivial matter), let us imagine that Definition 1 were replaced by the statement, 'The mass of a body is a measure of it proportional to its weight.' Assuming mass to be a primary quantity, we should then be at liberty to make an arbitrary choice of standard, assigning unit mass to some body fabricated with particular care and kept for purposes of ultimate reference, and by use of the beam balance and sets of subsidiary standards (or 'weights') be in a position to assign a mass-value to any body whatever (see p. 132).

At first sight, the procedure that we have indicated would appear to be commendably direct and altogether unexceptionable. However, a little consideration will show that this favourable verdict is essentially mistaken. The 'very accurate experiments with pendulums', to which Newton referred, merely carried Galileo's conclusion one stage farther – namely, the conclusion that the acceleration due to gravity, at any point on the earth's surface, is the same for all bodies, so far as can be experimentally ascertained. (In 1922 Eötvös greatly improved on Newton's accuracy, when he failed to detect, by indirect means, any difference in this acceleration greater than six parts in 10^9 among the bodies that he investigated.) Now this conclusion, regarding constancy of acceleration, has no necessary relevance to the mass concept: to give it that relevance we have to involve the force concept as well. We have to see the acceleration as due to the action of the weight of the body, to recognize the weight as a force, and effectively to accept the relationship between the measures of acceleration and force which is represented by the formal equation $F = ma$ (see p. 101). If we do this, then, of course, masses may be determined in terms of weights – but, in that case, we shall have committed ourselves irrevocably to a particular definition of force. We should like to avoid the appearance of such premature committal – after all, the force of gravity is only one of the forces effective in the world; we should like to have some information regarding the effects of other forces before finalizing our conceptual scheme. There is also another, more subtle, point of criticism into which we cannot enter here, but to which we shall later briefly refer (see p. 144).

If we regard the criticism of the last paragraph as valid, then we must necessarily conclude (i) that there is no quantifying definition of mass amongst Newton's eight 'Definitiones', or in his gloss on Definition 1, (ii) that, once such a definition of mass is framed, Newton's Definition 1 becomes an acceptable definition of density, (iii) that, under the same condition, Definition 2 becomes an acceptable definition of momentum.

Let us turn now to Law 2. First, we make a trivial point: Law 1 is merely a special case of Law 2. If there is no 'force impressed', then, according to Law 2, there is no change of motion (momentum), and therefore no change of velocity – and that is the essence of Law 1. This consideration reinforces our conclusion that Law 1 is redundant. Secondly, we draw attention to the disconcerting abruptness of the language of Law 2. Time is not mentioned (and the statement is not referred to 'a body', as are the statements of Laws 1 and 3). Possibly the omission of the time factor reflects Newton's preoccupation with collision phenomena in writing this law, though the law clearly refers to 'impressed forces', and in his gloss on Definition 4 Newton wrote 'An impressed force may arise in various ways, as from a blow, a pressure, a centripetal force.' It must be held, then, in the absence of any precise statement to the contrary, that forces other than those of impact are intended in Law 2. Newton's expositors have also held the view that the time factor is a necessary ingredient of the law, though they have produced no convincing explanation of Newton's omission to mention it. They have consistently rewritten the law (in modern phraseology), 'Rate of change of momentum is proportional to the moving force impressed, and takes place in the straight line in which that force is impressed.' We may find support for this change of wording in the form of Definition 8. This definition, in reference to one particular type of impressed force (see above), states, 'The motive quantity of a centripetal force is a measure of it proportional to the motion [momentum] which it generates in unit time.' Indeed, if we compare Definition 8 with the modified version of Law 2, we realize at once that there is a considerable overlap in content between these statements – the latter admittedly more general than the former (and saying something about the vectorial character of force and momentum) – but not, it would appear, sufficient difference in structure for the one to be classified as law, the other as definition.

In its modified (as in its original) form, Law 2 is not as precisely quantitative as are Definitions 1 and 2. There is a constant of proportionality which remains unspecified. In the representative equation of the quantifying definition of a derived quantity there should be no such unspecified constant subject to arbitrary choice. The lack of (numerical) precision in Law 2 may possibly indicate that Newton was vaguely aware that somehow in this statement a new primary (rather than a derived) quantity was being introduced. When we examine the matter, however, we find the statement, on its own, to be quite inadequate for this purpose. In our survey, so far, we have failed to discover an acceptable quantitative definition of mass – or an independent quantitative definition of force – and in Law 2 we find the measures of each of these quantities involved, one on one side of the representative equation ($\dot{p} = m\dot{v} = kF$, with k as the unspecified constant), and one on the other. Law 2 cannot serve as a definition of both.

We have now traversed all the 'quantifying recipes' that we identified in the eleven statements of Newton (with the exception of Definition 7, which assuredly has nothing to contribute to the solution of our difficulties), and we have not found in any of them the content that we were expecting to find. Possibly, in the process, we have clarified our own views as to what precisely it was that we were looking for, but that has been our only gain, hitherto. What we were expecting to discover was some statement of a law of nature – an uncovenanted regularity discerned in the kinematical behaviour of systems of bodies, and expressed in purely kinematical terms – into the expression of which we could bring 'understanding' by the importation of some non-kinematic quantity, suitably defined. Ultimately, on this basis, we hoped, in particular, to develop a definition of 'force', in scientific terms, which would not be wholly incompatible with the ordinary meaning conveyed by that word in everyday use (see p. 98). So far, the nearest that we have come to 'an uncovenanted regularity ... expressed in purely kinematic terms' has been the empirical result that the acceleration due to gravity is constant for all bodies at a given place, but for reasons that we have given, we have been unable to build on that regularity in the way that we hoped. Hence arises our interim assessment of failure.

At this stage we have only Law 3 in which to discover the content that we have hitherto missed. Law 3 is not a quantifying statement in the regular sense: it merely identifies an equality. On the other hand it has the appearance of making an assertion about the real world: action and reaction (on any arbitrary definition of these quantities) need not necessarily be of the same magnitude. To understand Law 3 we must, of course, interpret it first on Newton's own terms. It appears to consist of a general statement, and a parallel statement of less generality. Let us begin with the less-general statement: 'The mutual actions of two bodies are always equal and act in opposite directions.' We remember Definition 4: 'An impressed force is an action....' Here, then, with some justification, we may rewrite the second statement of Law 3, without changing its meaning, in the more modern form, 'Whenever two bodies interact, the mutual forces between the bodies are equal in magnitude and are oppositely directed.' (This is precisely the novel assertion that we ascribed to Newton – in a more restricted context – when we were considering the collision experiments in section 5.4.) Now, if we take in Law 2 (as Newton's expositors amended it – see p. 107), we justify the further development: 'Whenever two bodies interact, the rates of change of the momenta of the bodies are equal, with opposite directions.' Finally, incorporating Definition 2, we have the form, 'Whenever two bodies interact, the ratio of the instantaneous rates of change of the linear velocities of the bodies (that is, the ratio of their linear accelerations) is constant in magnitude (the mutual accelerations being oppositely directed); this numerical ratio is the inverse ratio of the masses of the bodies.' Here, at last, for the particular case of the isolated system of two bodies, we have a two-component statement of the type we have been looking for: the first component asserts a kinematical regularity grounded in experience, the second component is a quantitative definition of mass based on the acceptance of that regularity.

It will be realized that we have reached our last result by clearing away the incubus of pseudo-definition with which the Newtonian formulation is encumbered in order to reveal its direct contact with experiment. In so doing, we have identified mass as the one additional primary (non-kinematic) quantity involved in the scheme which Newton proposed. This identification securely made, we accept, at once, the Newtonian definition of force contained in Law 2 (setting $k = 1$ as is appropriate, since the statement of the 'law' is now the quantifying definition of a derived quantity). Indeed, now that we 'really understand' what we mean by mass and force (at least in the two-body situation) we are likely to revert, for sake of economy of words, to a form of expression of the (restricted) law of nature more nearly akin to Newton's original form. Let us say, then, understanding what we say, 'Whenever two bodies interact, the mutual forces between the bodies are equal in magnitude and oppositely directed.'

In a formal way we may now deal with the general statement of Law 3 quite briefly: 'An action is always opposed by an equal reaction.' Obviously, the word 'opposed', in this context, has no more than a directional connotation (it is the counterpart of 'in opposite directions' in the second part of the law). Once we have accepted that clarification, we may interpret the statement in a free paraphrase as follows: 'Forces never come into play singly, but always in pairs of equal and oppositely directed forces effective as between bodies (or particles).' When the general statement of the law is set out in this form, the need for the less general statement which Newton added to it vanishes completely: the general statement itself is sufficiently explicit. To this conclusion we need add only one comment: in respect of a many-body system, when any particular body is likely to be acted on by more than one force, the relevance of Newton's Corollary 1 (see pp. 103, 112) for a full analysis of the situation becomes obvious.

In a formal way, then, we have dealt with Law 3 in its entirety. We have expressed it in terms which, in principle at least, we now profess to understand and in a form which admits of no doubt that it is a statement about the real world – a 'law of nature' in the proper sense of the term. We might justifiably think that this is the conclusion of the matter. Because, however, this is the first law of nature that we have encountered in this book, and because in the Introduction (p. 14) we made the remark that the number of such laws is relatively small, we take this opportunity to consider it further. We have said that Newton's system of dynamics has been eminently successful. Even today it needs modification only in relation to very extreme conditions – when velocities comparable with the velocity of light are concerned, or distances smaller than the radii of atoms. The reason for that success is clearly contained in the fact that the law of nature on which it is based is well-nigh perfect in form as mirroring the form of the world. We may be curious to ask how Newton achieved this highly satisfactory result. It certainly was not achieved, by means of any logical process, from an exhaustive compilation of experimental observations. Even the restricted form of Law 3 referring to two-body systems could not have been deduced in that way (the adverbial 'whenever' in our version of its statement immediately excludes that possibility), to say nothing of the general form. Laws of nature cannot be validated absolutely on the basis of prior knowledge of natural phenomena, they must be formulated initially largely through processes of intuitive 'sympathy' or understanding. That was the way with Newton. Such laws can only be tested against subsequent observation – and be rejected or modified if they fail adequately to assimilate the new facts so disclosed. They are perpetually vulnerable in this way. Long after he had completed *Principia*, Newton wrote (*Opticks*, 1704), 'the arguing from experiments and observations by induction ... is the best way of arguing which the nature of things admits of ... if no exception occurs from phenomena, the conclusions may be pronounced generally. But if at any time afterwards any exception shall occur from experiments, it may then begin to be pronounced with such exceptions as occur.' Eventually, when he came to write the 'General scholium' for the second edition of his major work (1713) he was more specific: 'In this philosophy particular propositions are inferred from the phenomena, and afterwards rendered general by induction. Thus it was that the ... impulsive force of bodies, and the laws of motion ... were discovered.' In this section we shall allow Newton the last word.

6.4   A revised formulation (I)

As a result of the detailed discussions of the last section we have reached the conclusion that Newton's formulation of the laws of motion, whilst unexceptionable in relation to content, is, by modern critical standards, unacceptable in the matter of presentation. Although, towards the end of that discussion, we were able to disentangle the threads of definition and law which are intimately interwoven in the original, the end result of our exercise is anything but tidy. We may now understand Newton; however, what we more urgently require is a neat formulation of laws and definitions in the modern idiom which will serve us for future reference and use. It is the object of this section to provide and discuss such a revised formulation. We do so forthwith, assuming familiarity only with kinematical concepts and quantities previously defined. In setting out our new formulation we include, in parenthesis, in relation to each formal statement, an indication of its status in the over-all scheme, and we enclose in square brackets such of our assertions as are no more than logical consequences of statements previously made.

Law 1 (qualitative and general). It is possible to assign to every identifiable portion of matter in the universe a unique mass-measure which is a positive number (when expressed in relation to the unit mass-measure arbitrarily assigned to a specified standard portion of matter).

Definition 1 (formal). A particle is an identifiable portion of matter which may be regarded as structureless for the purpose in hand; a body is an identifiable portion of matter which must be regarded as constituted of discrete particles.

Definition 2 (quantitative and special). When two particles form an isolated system, the inverse ratio of the measures of the instantaneous accelerations of the particles is in all circumstances the same as the ratio of the mass-measures of the particles.

Definition 3 (quantitative and general). Whenever a particle of which the mass-measure is m has acceleration of measure a, the particle is said to be subject to force of measure (F) equal to ma, the direction of the force being that of the acceleration.

[Because acceleration is a vector quantity, and is therefore subject to the rules of vector resolution, it may correspondingly be said that in a given situation the (total) force acting on a particle has components as given by those rules.]

Law 2 (quantitative and general). When N particles form an isolated system (however large the value of N), there is at every instant a unique way of resolving the accelerations of the particles, each into $N-1$ components, so as to produce $\frac{1}{2}N(N-1)$ pairs of oppositely directed components (the two members of any pair relating to different particles). The properties of this mode of resolution are (i) that the values of the $\frac{1}{2}N(N-1)$ ratios of the measures of the paired component accelerations remain constant throughout time, (ii) these $\frac{1}{2}N(N-1)$ ratios are not all numerically independent – they reduce, in N different ways, to $N-1$ ratios the values of which are independent each of all the others, (iii) for any one of these N ways of reduction, the $N-1$ independent ratios are the ratios of the $N-1$ component accelerations of one particular particle to the component accelerations, one referring to each of the other particles, which are paired with them.

Definition 4 (quantitative and general). The $N-1$ independent ratios obtained in any of the N ways of reduction and referred to in Law 2(iii) are the same as the ratios of the mass-measures of the other particles to the mass-measure of the particular particle all of whose acceleration components are represented in those ratios.

[When mass-measures have been assigned to all the N particles referred to in Law 2, in terms of unit mass-measure arbitrarily assigned to one of them as standard, on the basis of Definition 4, then the statement of the law will be seen to be equivalent to the shorter statement: 'When N particles form an isolated system (however large the value of N), there is at every instant a unique way of resolving the total forces acting on the particles, each into $N-1$ components, so as to produce $\frac{1}{2}N(N-1)$ pairs of oppositely directed component forces (the two members of any pair relating to different particles). The properties of this mode of resolution are that for each of the $\frac{1}{2}N(N-1)$ component-force pairs the associated component forces are equal in magnitude.']

We make the following two comments concerning this set of four definitions and two laws. First, except in the qualitative assertion of Law 1, these statements refer exclusively to particles. It may be – indeed, it might appear that it should be – the case that a system, which is in every way adequate for the description of the dynamics of particles, must contain all that is necessary for the subsequent development of the dynamics of bodies constituted of particles. However, that is a question which we cannot lightly regard as predetermined; we shall have to discuss it in detail as we develop our argument. We might remark, as a matter of history, that the first person to criticize Newton's formulation, in the general way that we have criticized it in the last section, was Ernst Mach (1838–1916), in a treatise published in 1883. Mach put forward an alternative formulation similar to ours, in that his fundamental law was expressed solely in kinematical terms. However, Mach's formal statements related to bodies (as Newton's had done). In this connexion we recall our earlier conclusion (p. 68), that in relation to the instantaneous state of motion of an extended object (solid figure), whilst it is possible to give an unambiguous specification of its velocity of rotation, it is impossible, without further consideration, to give a similar specification of a unique velocity of translation. For this reason we have avoided speaking of bodies in our basic definitions and in Law 2. In so far as Law 1 refers to bodies, therefore, it is looking ahead to future developments: to that extent it is extraneous to the limited object of providing a sufficient basis for the dynamics of particles exclusively.

Our reason for giving Law 1 this additional content – and this is our second comment – is twofold. We wish at the outset to state our conviction that a complete practical system of dynamics can be achieved by the route we are taking, and we wish to provide a peg on which to hang some of our later definitions. Thus, Definition 1 has a less arbitrary aspect if the phrase 'identifiable portion of matter' has previously been used – and in Definition 2 we are able to avoid mention of the directional relation of the accelerations (merely having regard to their measures, irrespective of direction), and of the 'constancy throughout time' of the ratio of those measures, once it has been firmly stated (as is deducible from Law 1) that a unique mass-measure ratio results in such circumstances. If we had found it necessary to include these details in the statement relating to the two-particle system, that statement would have taken on the character of a law rather than a definition (the details, in fact, are given specifically by Law 2, if we interpret it for the special case $N = 2$).

6.5   A revised formulation (II): the additivity of mass

In the last section we put forward a revised formulation of Newton's laws of motion, having reference exclusively to the kinetics of particles. Assuming that formulation to be satisfactory for its immediate purpose, we left open the question whether in fact it provides sufficient basis for a complete system of dynamics in which the kinetics of bodies is naturally included. In this section we take the first step towards an answer to that question. We examine the proposition, 'The mass of any body is equal to the sum of the masses of its constituent particles,' to see whether this proposition can be deduced unambiguously from the definitions and laws that we have proposed. If it cannot be so deduced, we shall obviously have very grave doubts regarding the validity of the 'extraneous' claim that we hopefully made in Law 1.

Let us first establish a general result which is a variant and extension of the simple parallelogram construction for the composition (or resolution) of vector quantities (see p. 23). Let $\overrightarrow{AB}$, $\overrightarrow{AC}$ (Figure 16) represent the vector quantities r_1, r_2: our problem is to devise a construction giving the representation, in the same diagram and on the same scale, of the vector quantity $m_1 r_1 + m_2 r_2$. Here m_1, m_2 are pure numbers – or the measures of scalar quantities of the same kind. Let B, C be joined, and let D be a point in BC.

$$\text{Then} \quad \overrightarrow{AB} = \overrightarrow{AD} + \overrightarrow{DB},$$
$$\overrightarrow{AC} = \overrightarrow{AD} + \overrightarrow{DC},$$

and, if $m_1 \overrightarrow{BD} = m_2 \overrightarrow{DC}$,   **6.7**

for any values of m_1 and m_2,

$$m_1 \overrightarrow{AB} + m_2 \overrightarrow{AC} = (m_1 + m_2)\overrightarrow{AD}. \qquad \textbf{6.8}$$

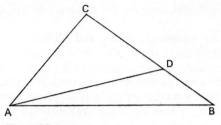

Figure 16

Obviously, the required construction is specified by equations **6.7** and **6.8**: if the point D is found which divides BC internally in the inverse ratio of the 'weighting factors' m_1 and m_2 (equation **6.7**), then the quantity $m_1\,\mathbf{r}_1 + m_2\,\mathbf{r}_2$ is represented by $(m_1 + m_2)\overrightarrow{\text{AD}}$ (equation **6.8**).

It is no accident that, in deriving the general result that we have just obtained, we used the symbol m to denote the 'weighting factor' in our problem. If, indeed, m_1 and m_2 are interpreted as the mass-measures of two particles situated at B and C, respectively, then equation **6.7** specifies the construction giving the point D such that $m_1\,\mathbf{r}_1' + m_2\,\mathbf{r}_2' = 0$, $\mathbf{r}_1'$ and $\mathbf{r}_2'$ being the displacement vectors specifying the positions of the two particles with respect to D. Moreover, equation **6.8** shows that for any (other) point, such as A,

$m_1\,\mathbf{r}_1 + m_2\,\mathbf{r}_2 = (m_1 + m_2)\bar{\mathbf{r}},$

$\mathbf{r}_1$, $\mathbf{r}_2$ and $\bar{\mathbf{r}}$ being the displacement vectors specifying the positions of the two particles and of the point D with respect to that other point. Because of the significance of these simple results, the point D is given a special name in this connexion; it is referred to as the *centre of mass* of the particles. The notion can be generalized for any number of particles: for any system of particles there is a unique point, the centre of mass of the particles, for which $\sum_s m_s\,\mathbf{r}_s = 0$.

In the guise of the problem of the *centre of gravity* (see p. 145) – and long before the invention of vector algebra – this result was effectively established by John Wallis, and published in his treatise of 1671 (see p. 93).

We return now to our problem of the additivity of mass. Let us consider three particles of mass m_0, m_1, m_2, respectively, situated instantaneously at A, B and C (Figure 16). Then if, in terms of Law 2, the significant pairs of oppositely directed component accelerations in this case are $({}_0\mathbf{a}_1, {}_1\mathbf{a}_0)$, $({}_1\mathbf{a}_2, {}_2\mathbf{a}_1)$ and $({}_2\mathbf{a}_0, {}_0\mathbf{a}_2)$, in obvious notation, we have

$$\left. \begin{aligned} m_0\,{}_0\mathbf{a}_1 &= -m_1\,{}_1\mathbf{a}_0, \\ m_1\,{}_1\mathbf{a}_2 &= -m_2\,{}_2\mathbf{a}_1, \\ m_2\,{}_2\mathbf{a}_0 &= -m_0\,{}_0\mathbf{a}_2. \end{aligned} \right\} \qquad \textbf{6.9}$$

From the first and third of equations **6.9** we obtain the result

$$m_0({}_0\mathbf{a}_1 + {}_0\mathbf{a}_2) = -(m_1\,{}_1\mathbf{a}_0 + m_2\,{}_2\mathbf{a}_0),$$

and, by incorporating the equality represented by the second of these equations, the alternative form

$$m_0({}_0\mathbf{a}_1 + {}_0\mathbf{a}_2) = -m_1({}_1\mathbf{a}_2 + {}_1\mathbf{a}_0) - m_2({}_2\mathbf{a}_0 + {}_2\mathbf{a}_1).$$ **6.10**

Suppose that we represent by $\mathbf{a}_0$, $\mathbf{a}_1$, $\mathbf{a}_2$ the 'total' accelerations of the three particles. We then have

$$\mathbf{a}_0 = {}_0\mathbf{a}_1 + {}_0\mathbf{a}_2,$$

and correspondingly for the other particles. On this basis, equation **6.10** becomes

$$m_0\,\mathbf{a}_0 = -m_1\,\mathbf{a}_1 - m_2\,\mathbf{a}_2;$$ **6.11a**

or, alternatively, $\quad m_0\,\mathbf{a}_0 + m_1\,\mathbf{a}_1 + m_2\,\mathbf{a}_2 = 0.$ **6.11b**

Let us now consider the instantaneous acceleration of D, the centre of mass of the particles at B and C. If $\mathbf{r}_1$, $\mathbf{r}_2$ and $\bar{\mathbf{r}}$ are the displacement vectors specifying the instantaneous positions of B, C and D with respect to any arbitrarily chosen origin, and if we denote the instantaneous acceleration of D by $\mathbf{a}_{12}$, we have

$$\mathbf{a}_{12} \equiv \ddot{\bar{\mathbf{r}}},$$

and (see above) $\quad (m_1 + m_2)\ddot{\bar{\mathbf{r}}} = m_1\,\mathbf{r}_1 + m_2\,\mathbf{r}_2.$

Thus $\quad (m_1 + m_2)\mathbf{a}_{12} = m_1\,\ddot{\mathbf{r}}_1 + m_2\,\ddot{\mathbf{r}}_2.$

But, in terms of our previous notation,

$$\ddot{\mathbf{r}}_1 \equiv \mathbf{a}_1, \qquad \ddot{\mathbf{r}}_2 \equiv \mathbf{a}_2.$$

Hence, we have $\quad (m_1 + m_2)\mathbf{a}_{12} = m_1\,\mathbf{a}_1 + m_2\,\mathbf{a}_2.$ **6.12**

Finally, comparing equations **6.11a** and **6.12**, we conclude that

$$m_0\,\mathbf{a}_0 = -(m_1 + m_2)\mathbf{a}_{12}.$$ **6.13**

We are in a position now to summarize our formal argument and to interpret its final conclusion in words. We started with an isolated system of three particles. Equations **6.9** may be regarded as exhibiting the quantitative basis, which is provided by Law 2, for the assignment of mass-measures to two of the particles (1 and 2) in terms of the mass-measure of the third (particle 0) taken as standard of reference. Equation **6.13** indicates that, as all three particles move (under the influence of their mutual forces – of whatever nature these may be), the acceleration of the centre of mass of particles 1 and 2 and the (total) acceleration of particle 0 are always oppositely directed and in constant ratio. If that ratio were used to define the mass-measure of a fictitious particle moving with the centre of mass of particles 1 and 2, the mass-measure of the particle would be $m_1 + m_2$ on the mass scale that we are using In other words, the mass of this 'fictitious' particle would be precisely equal to the sum of the masses of the two 'real' particles which it 'represents'.

The result that we have just formulated is valid for any isolated system of three particles, whether these remain permanently separate from one another or whether two of them (particles 1 and 2 of our formalism) are permanently associated so as to constitute a composite entity (a rudimentary 'body' according to our Definition 1). In that case, so far as the motion of the centre of mass of this composite entity is concerned, this association of two particles behaves as if characterized by a mass equal to the sum of the masses of its constituents.

Obviously, we may repeat our argument indefinitely. At any stage, in relation to Figure 16, we assume that we have a single particle of mass m_0 at A, a composite entity of mass $\sum\limits_{k=1}^{s} m_k$ of which the centre of mass is at B, and a single particle of mass m_{s+1} at C. In relation to the particle at C, as well as in relation to the 'standard' particle at A, the composite entity behaves (so far as the motion of its centre of mass is concerned) as if it were a particle of mass $\sum\limits_{k=1}^{s} m_k$ situated at B. The argument, then, proceeds as we have given it, and we conclude that if, in fact, the particle at C were permanently associated with the original composite entity of s particles to form a composite entity of $s+1$ particles, this larger entity would behave (in relation to the motion of its own centre of mass) as a particle of mass $\sum\limits_{k=1}^{s+1} m_k$ situated at that particular (moving) point.

Here, we are very close to establishing the proposition that we set out to examine – 'The mass of any body is equal to the sum of the masses of its constituent particles' – using only the definitions and laws relating to the kinetics of particles as we formulated them in section 6.4. We should not be too hasty, however, to regard our investigation as effectively concluded at this stage, for a little consideration will show that we have obtained a result that is more detailed and significant than perhaps we had anticipated. When our 'composite entity' consisting of a very large number of permanently associated particles has taken on the more familiar aspect of a solid body of well-defined size and shape (though we do not deny that the constituent particles may still be in small-amplitude motion within it) we have an object such that the centre of mass of its constituent particles is a permanently identifiable point within its figure. What our analysis shows is that such a body, under the action of an 'external force' (namely the resultant of the forces acting from the 'standard particle' of our argument on all the particles in the body), at least in relation to its purely translational motion, moves as if all its particles were concentrated at the centre of mass, all the forces (forces both 'internal' and 'external', in relation to the whole body) being 'transferred' with the particles, only the directions of action of these forces being kept constant. In this transference, we notice, all the internal forces cancel in pairs (p. 112), only the external forces are left as determining the motion.

In assessing this last detailed result, it will be seen that we have, indeed, derived the proposition that we quoted at the outset of this section directly from our system of definitions and laws, but that in so doing we have revealed a hidden aspect of imprecision in the form of the proposition itself. The mass concept in relation to an extended body has precise meaning, so it appears, only when it is understood within a context in which the concept of centre of mass has already been defined. This is a gain in understanding; moreover, the result from which it derives provides, also, a feature which has hitherto been found missing in our investigations. Kinematically, there is no way of assigning a uniquely significant translational velocity (or acceleration) to an extended body (pp. 68, 112): kinetically, it now appears, the velocity (or acceleration) of the centre of mass of the body is a uniquely significant vector quantity. With the acceptance of this fact, and the additivity of mass, which depends on that acceptance in the rather subtle way that we have suggested, together with the theorem concerning the effect of an 'externally applied' force which we have established in the course of our argument (see above), we have a secure basis for the development of the kinetics of translation of ideal rigid bodies, assuming no more than the theoretical basis of the kinetics of particles which we formulated in section 6.4. In the next section we must look into the problem of rotational motion. We must attempt to see whether that aspect of the total motion of bodies, also, can be dealt with coherently without the introduction of any further hypothesis or law of nature into our theoretical scheme.

6.6 A revised formulation (III): rotational motion

The problem of rotational motion, on any fundamental level of discussion, is very much more complicated than that of translational motion: for the purposes of this book we shall have to be content with a detailed consideration of certain very simple cases, supported, where appropriate, by the mere statement of generalizations that we shall not always be able to justify in full.

Let us consider first the case of an isolated body rotating with constant angular speed ω about an axis passing through its centre of mass and having a fixed direction in space. Because the body is isolated, by definition it is uninfluenced by any external force, thus the linear acceleration of the centre of mass is zero. If the velocity of the centre of mass is also zero, the linear speed of any constituent particle of the body which is situated at a distance q from the axis of rotation is $q\omega$, and the acceleration of this particle is of magnitude $q\omega^2$ and is directed towards the axis of rotation. If the particle that we are considering is of mass m, the resultant of all the internal forces acting on it must be of magnitude $mq\omega^2$ and must also be directed towards the axis of rotation.

The resultant forces that we have just identified increase, throughout the body, from zero when the body is completely at rest, in proportion as the square of the angular velocity. In any real body an increasing 'imbalance' of the internal forces acting on each particle of the body can arise only from a progressive change in the over-all configuration of the particles, that is, from an increasing 'deformation' of the body, as its angular velocity increases. The equal and oppositely directed forces of mutual interaction between any two particles must depend, as to magnitude, on the separation of the particles.

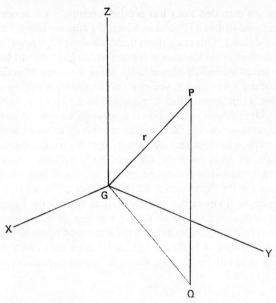

Figure 17

The equilibrium configuration of particles obtaining when a body is completely at rest is such that the forces acting on each single particle cancel completely: when the body is rotating, the equilibrium configuration of particles changes until a resultant force of appropriate magnitude acts on each particle towards the axis of rotation. Obviously, here, we are again concerned, as we were when we were discussing the phenomena of impact in physical terms (p. 95), with the *elastic* properties of bodies. Possibly, we have carried that discussion a little farther in this case.

Let us summarize our discussion of this first simple case, before we proceed. If this is our introduction to the kinetics of rotation, it will appear at the outset that we have to concede that no real understanding of even the simplest case is possible without our abandoning the concept of the 'ideal rigid body' (see p. 84) to which tradition accords such a prominent place in this particular field of inquiry. At the level of fundamental discussion, this is no trivial remark: no full understanding of rotational motion in the real world can ever be reached in terms of the rigid-body concept, but it is an extraordinarily useful concept, none the less. From now on, taking refuge in the reflection that on the macroscopic scale the elastic deformations are relatively very small, we shall generally forget about them: unless we decide otherwise in pursuit of particular enlightenment, our 'experimental' bodies will be ideally rigid bodies, exclusively. Let fly-wheels shatter, when elastic limits are exceeded!

In the case that we have just been considering, that of an (ideal rigid) body rotating steadily about an axis through its centre of mass, it is a simple matter to show that transference of the (total) forces instantaneously acting on all the particles of the body, with their directions unchanged, to the centre of mass as effective point of application, produces a resultant which is zero – as must necessarily be the case if these forces are entirely of 'internal' origin. Let GX, GY, GZ (Figure 17) be rectangular axes through G, the centre of mass of the body, GZ being the axis of rotation. Let P be the position of the constituent particle of mass m distant $\mathbf{r}$ from G. If PQ is drawn perpendicular to the plane XGY, obviously, in terms of our previous notation, $GQ = q$. Now, because G is the centre of mass,

$$\sum m\mathbf{r} = 0. \qquad \textbf{6.14}$$

Also $\mathbf{r} = \mathbf{q} + \mathbf{z},$

where $QP = z$. However, q and z, being two of the (cylindrical) coordinates of the position of the particle, are entirely independent one of the other. This being the case, equation **6.14** cannot be valid unless, independently,

$$\begin{aligned}\sum m\mathbf{q} &= 0, \\ \sum mz &= 0.\end{aligned} \qquad \textbf{6.15}$$

(In the first of equations **6.15**, q is a vector in two dimensions; in the second z is effectively a scalar quantity which can have both positive and negative values.) From the first of these equations we obtain

$$-\omega^2 \sum m\mathbf{q} = 0. \qquad \textbf{6.16}$$

In this equation ω is a constant quantity, and in the context of our problem its vectorial character is of no significance. The left-hand member of the equation represents the vector sum of all the forces acting (inwards towards the axis) on the particles of the body 'because of the rotation', the forces being considered as transferred to a common point. Equation **6.16**, in fact, establishes the result that we set out to prove.

Let us digress briefly at this stage to consider a very simple extension of the result that we have just obtained. We wish to show that an isolated body cannot continue to rotate with constant angular velocity about an axis fixed in space and in the body and which does not contain its centre of mass. Let rectangular axes OX, OY, OZ (Figure 18) be taken, the axis of Z, as before, being the axis of rotation. Let the centre of mass of the body be situated at G in the plane XOY. The other features of the figure have the same relation to OZ as the corresponding features had to GZ in Figure 17. In particular, $OQ = q$, $QP = z$; also $GP = r$, as in the previous case. Let $OG = \bar{q}$.

Then $\mathbf{r} = \mathbf{q} - \bar{\mathbf{q}} + \mathbf{z},$

and, as before, $\sum m\mathbf{r} = \sum mz = 0.$

Thus $$\sum m\mathbf{q} = \bar{\mathbf{q}} \sum m,$$

and, in consequence, $-\omega^2 \sum m\mathbf{q} = -\omega^2\bar{\mathbf{q}} \sum m. \qquad \textbf{6.17}$

119   A Revised Formulation (III): Rotational Motion

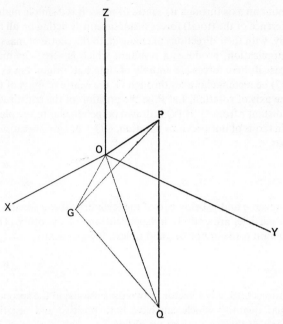

Figure 18

Equation **6.17** shows that if the (total) forces acting (inwards towards O Z) on all the particles of the body in this case were transferred to G their resultant would be along $\overrightarrow{GO}$ and of magnitude equal to $\omega^2 \bar{q} \sum m$. Because this resultant is of finite magnitude, solely internal forces cannot be responsible for it. If these were the only forces available, the motion specified could not persist. We put forward a negative proposition for consideration and we have validated it: it is interesting to note that in so doing we have shown that the 'external' force system which is necessary to sustain the motion that we specified is one which, if all the forces acting on the individual particles of the body were transferred to the centre of mass, and all the mass of the body were concentrated there, would cause this point to move according to the specification of the problem, with constant angular speed ω about O Z. Of necessity, of course, such a force system would have to rotate with the body.

Suppose now – this is our second case for detailed consideration – that an ideal rigid body is instantaneously rotating about an axis through its centre of mass with angular speed ω, and that at the same instant its angular acceleration is α about the same axis. The centre of mass is at rest and unaccelerated. We consider a constituent particle of the body of mass m, as before. The components of linear acceleration of this particle are $q\omega^2$ towards the axis of rotation and $q\alpha$ along the tangent to the circular path (of radius q) in which the particle moves. The components of the total force acting on the particle are $mq\omega^2$ and $mq\alpha$, respectively, in the directions of the accelerations. We have just shown that the components proportional to ω^2 would have zero resultant if transferred unchanged in direction to the centre of mass. A precisely similar result applies to the components proportional to α. The vectorial representation of this component, for the particle of mass m, is $jmq\alpha$ (here the symbol j represents the rotation of the two-dimensional vectorial direction through $\frac{1}{2}\pi$ radians in the positive sense), so the resultant in question is $j\alpha \sum mq$. If $\omega^2 \sum mq = 0$, then $j\alpha \sum mq = 0$, also.

The result that we have just derived is not immediately revealing. It is consistent with the assumption that the total forces acting on the constituent particles of the body arise solely from the internal interactions of those particles one with another (this was the conclusion that we reached and accepted in the first case that we studied). However, we cannot lightly accept a similar interpretation in the present case. The macroscopic motion of the body is accelerated motion, and it is contrary to our general understanding of the principle of inertia that this could be so if there were no external forces acting on the body. We are led, then, to the only possible alternative assumption, namely, that external forces are effective, but are of such a character that if they were transferred, from the particles on which they act, to the centre of mass of the body, they would also be found to have zero resultant, as the internal forces of necessity must have. Indeed, if we had not wished to explore the matter formally and in detail, we might have reached this conclusion directly, by a simple verbal argument, given our postulated conditions (that the centre of mass of the body is unaccelerated, whereas its rotational motion is accelerated) and our intuitive notions concerning inertia generally.

With the acceptance of this alternative interpretation, we now know how we must approach the problem under discussion – but the problem is not thereby solved, it is merely prepared for solution. Our aim is to devise a recipe, which is consistent with the laws of particle kinetics as we have stated them, whereby the angular acceleration of a body (about a particular axis through its centre of mass) can be calculated in terms of the detailed specification of a system of external forces which produces no acceleration in the centre of mass of the body. Obviously, in such a recipe there must be some parameter which is representative of an inertial attribute of the body appropriate to the situation in question.

Our problem having been clarified in this way, we note that attention has been focused on the lines of action, or points of application, of the external forces, rather than on their directions exclusively. For the purpose of determining the translational motion of a body, the external forces are effectively regarded as 'non-localized' vectors (in that they are transferable to the centre of mass of the body): in respect of the rotational motion, we must regard them as 'localized' vectors. When rotation about a particular axis is involved (in which case any component of force parallel to that axis is clearly of no concern to us), we are thus led to consider the possibility that the perpendicular distance from the axis to the line of action of a force (or force component) in a plane at right angles to the axis is a significant quantity in relation to the rotational effect of that force. Let us examine that possibility forthwith, for the case that we are considering. For brevity of description we shall refer to the product of $F_\perp$, the measure of a force component in a plane at right angles to the axis of rotation, and q, the measure of the perpendicular distance of the line of action of $F_\perp$ from the axis, as the measure of the *moment of the force* about the axis ($G = F_\perp q$). Then, for the force components $mq\alpha$ of our present problem (these are the force components 'responsible for' the angular acceleration), the sum of the moments of these components about the axis of rotation is given by

$$G = \sum mq^2\alpha,$$
$$\text{or} \quad G = \alpha \sum mq^2. \qquad\qquad 6.18$$

We shall later have to discuss the vectorial character of our new concept (the moment of a force); for the present, equation 6.18 is merely a relation between measures of quantities, without explicit vectorial significance.

The first feature of interest in equation 6.18 – and it is very definitely an encouraging feature – is that the quantity $\sum mq^2$ is a parameter characteristic of the distribution of the mass of the body about the axis of rotation and having 'inertial significance'. We may interpret the equation qualitatively in words: the measure of the angular acceleration is determined by the sum of the moments, about the axis of rotation, of all the forces acting on the individual particles of the body and effective about that axis, and an inertial parameter characteristic of the distribution of the mass of the body about that axis. Again, for sake of brevity, let us refer to the inertial parameter of which the measure is $\sum mq^2$ as the *moment of inertia* of the body about the axis concerned.

The second point of interest in equation 6.18 is that as it stands it refers to the 'effective' components of all the forces – both internal and external forces – which act on the individual particles of the body. We are looking for a recipe which refers to the external forces only. If we can justify the conclusion that the sum of the moments of the internal forces is zero (about any possible axis of rotation of the body) we may interpret the quantity G in equation 6.18 as simply the sum of the moments of all the effective components of the external forces concerned. In that case we shall have

$$\sum {}_eF_\perp q' = \alpha I, \qquad\qquad 6.19$$

where ${}_eF_\perp$ is the measure of an external force component in a plane at right angles to the axis of rotation, q' is the measure of the perpendicular distance of the line of action of this component from the axis, and I is the measure of the moment of inertia of the body about that axis.

Let it be said at once that when faced with the problem of 'devising a recipe', as we are in our present problem (see p. 121), without any very clear guidance concerning where to start, we should not, in an expository account of an established science, start otherwise than along the route which in fact has already been shown to lead to success. That is what we have done, with hindsight, in this case. At the outset we introduced the concept 'moment of a force', almost arbitrarily as it might appear – as if this were the way in which the concept originally arose in the actual development of the science of dynamics. For sake of historical truth, we should make it clear before we proceed farther, that the facts are otherwise. As a result of his study of the lever, Archimedes had already convinced himself, in the third century B.C., that the turning effort of a load is proportional to the length of the arm across which it acts, and this 'empirical' result had been generalized by Leonardo da Vinci (1452–1519) almost two hundred years before the concept of force was finally given a dynamical basis by Newton. We may have to confess that the full development of the concept 'moment of a force' was slow to mature (it was finally codified in its dynamical aspect by Louis Poinsot (1777–1859) in 1834), but there can be no suggestion that the theorists of the early days were 'without any very clear guidance concerning where to start' in relation to the problem of accelerated rotational motion. The rudimentary notion of the moment of a force had been available to them from pre-Newtonian times.

In the light of our last remarks, let it be stated categorically that equation 6.19 represents the universally accepted 'successful' solution of the problem that we are presently considering. It remains, then, as an urgent necessity, for us to justify the assumption on which its validity depends – the assumption, that is, that the sum of the moments of the effective components of the internal forces acting on the constituent particles of any body is zero (about any possible axis of rotation of the body). This result does not follow directly from the laws of particle kinetics as we stated them in section 6.4. Effectively, we followed Newton in our formulation of Law 2 (p. 111), and when we finally came to paraphrase that law in terms of the force concept (p. 112) we had merely pairs of equal and oppositely directed forces acting as between each pair of particles (as Newton had in Law 3 (p. 103) between any two bodies) – without any further statement made concerning the precise directions of those forces, either in Newton's version or our own. Until now, no further qualifying statement has in fact been necessary. Now, however, such a further statement is required. Unless the mutual forces acting between any particle and any other particle in a composite body are not only equal, but also oppositely directed along a common line of action, it will not in general be the case that the sum of the moments of the effective components of these forces about any possible axis of rotation of the body is zero. If, then, we are to base the laws of kinetics of the rotation of bodies securely on the laws of kinetics of particles, as we wish to do, we must add to our earlier formulation the additional law:

Law 3. The equal and oppositely directed forces of interaction of any pair of particles are effective in the same straight line.

Obviously, we might, without undue complication, have written this

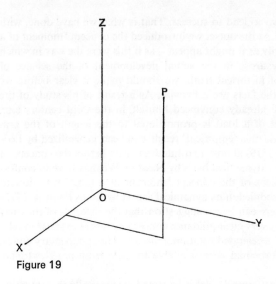

Figure 19

additional statement into our original Law 2. We omitted it for the very good reason that in practical terms the need for it becomes apparent only when the phenomena of rotational motion are involved. It is not unimportant that we should have a clear appreciation of the experimental basis of the various fundamental assertions that we find it necessary to make in constructing our theories. However that may be, we now add this third instalment of 'law' to our revised formulation of the system of Newtonian mechanics, and so justify equation **6.19** as the appropriate dynamical equation relating to the instantaneous situation postulated (p. 120) as the second simple case for our detailed consideration in this section.

Digressing from our discussion of the first case (p. 119), we drew attention to the fairly obvious fact that rotation with constant angular velocity about an axis other than through the centre of mass of a body cannot persist as steady-state motion in the absence of external forces. Before we leave our second case, it is necessary to refer to an analogous fact which is not so obvious, and which, indeed, it is beyond our competence to discuss in detail. The situation that we postulated in this case (constant angular acceleration about an axis through the centre of gravity) is not one that can be maintained indefinitely, the axis of rotation remaining fixed both with respect to the body and in space, even when an external (constant) force system is operative, except in special circumstances. For this state of affairs to persist, the axis of rotation must be one of the *principal axes of inertia* of the body. Though we cannot discuss this assertion in detail, and validate it as we did our cautionary assertion in the other case, it will be informative, at least, to define the new concept contained in it, for the general notions of *principal axes of inertia* and

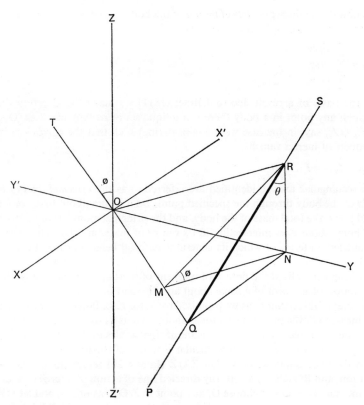

Figure 20

principal moments of inertia are fundamental for any more advanced treatment of rigid-body dynamics. It must be the aim of any elementary account to lay the foundations on which later developments are to be based.

We note, first of all, what is obvious from its mode of definition (p. 122), that the moment of inertia is not a single-valued quantity for any body. The magnitude of this quantity varies with the axis to which it refers: without reference to a specified axis, the term is meaningless. Suppose, then, that OX, OY, OZ (Figure 19) are rectangular axes fixed with respect to a rigid body, and that a representative particle of the body of mass m is situated at P (x, y, z). Let Let I_X, I_Y, I_Z be the moments of inertia of the body about the coordinate axes, the origin O being any point in the body. Then

$$I_X = \sum m(y^2 + z^2),$$
$$I_Y = \sum m(z^2 + x^2),$$
$$I_Z = \sum m(x^2 + y^2).$$

We now define three *products of inertia* of the body, corresponding to the axes that we have chosen, by the equations

$$I_{XY} = \sum mxy,$$
$$I_{YZ} = \sum myz,$$
$$I_{ZX} = \sum mzx.$$

On the basis of a result due to J. Binet (1811) we make the assertion that through any point in a body there is a unique set of rectangular axes (OX', OY', OZ', say, in the case we are considering) such that the corresponding products of inertia vanish

$$I_{X'Y'} = I_{Y'Z'} = I_{Z'X'} = 0.$$

The rectangular axes so identified are referred to as the *principal axes of inertia* of the body through the specified point. So it comes about that, for any rigid body, the total mass of the body, and the principal moments of inertia of the body about axes intersecting in its centre of mass, are the basic inertial parameters in terms of which its general dynamical behaviour may be evaluated.

Having generalized the notion of moment of inertia, let us now generalize the notion of moment of a force about an axis, and investigate the vectorial character of this quantity as we promised to do (p. 122). Previously we defined the measure of the moment of a force about an axis as the product of $F_\perp$, the measure of a force component in a plane at right angles to the axis of rotation, and q, the measure of the perpendicular distance of the line of action of $F_\perp$ from the axis. Suppose, now, that $Z'OZ$ (Figure 20) represents the axis of rotation, and PQRS the arbitrarily directed line of action of a force of which the measure is F. Let us choose O, any point in $Z'OZ$, as origin, and let OX, OY, OZ constitute a set of rectangular axes through O. Let PQRS intersect the coordinate planes XOY and YOZ in Q and R, respectively, and let RN be drawn parallel to ZO to intersect OY in N. Then RN, being perpendicular to the plane XOY, is perpendicular to QN, and if the length QR in the diagram is taken to be proportional to F, QN is proportional to $F_\perp$, the measure of the component of this force 'in a plane at right angles to the axis of rotation'. If $\angle QRN = \theta$, then $F_\perp = F \sin \theta$. Obviously this result is independent of our choice of the origin O, as is the value of q the measure of the perpendicular distance of O from QN, (q is essentially the perpendicular distance between $Z'OZ$ and the plane QRN which is parallel to $Z'OZ$). We have shown, then, for the general case, that our original definition specified a quantity which is independent of the plane at right angles to the axis in which the force is considered to be resolved, and that the measure of this quantity is $F \sin \theta . q$ in terms of the quantities that we have now defined. The measure of the moment, about any axis, of a force F, of which the direction is inclined at an angle θ to the axis, is $F \sin \theta . q$, where q is the perpendicular distance of the line of action of the force from the axis.

Let us now consider the matter from another point of view, regarding the point O as fixed, and the line PQRS as fixed in space, at a perpendicular distance d from O. Z'OZ, from this point of view, becomes an arbitrary direction through O, and OT, the direction through O at right angles to the plane OQR a 'fixed' direction specified by O and the line of action of the force F. In conformity with our previous definition, the measure of the moment of the force F about OT is given by Fd. In Figure 20 this measure is represented by twice the area of $\triangle$ OQR. On the same scale, the measure of the moment of F about Z'OZ is represented by twice the area of $\triangle$OQN. Let us represent these quantities by G_0 and G, respectively, and let $\angle$TOZ $= \phi$. Now, triangles OQR and OQN have common base OQ, and RN, the join of the apices of these triangles, is perpendicular to the plane XOY and therefore to OQ. This being the case, a plane containing RN may be found which is at right angles to OQ. Let this plane intersect OQ in M. Then MR and MN are perpendicular to OQ, $\angle$RMN $= \phi$, and $\angle$RNM $= \frac{1}{2}\pi$. As a result, we have

$$\frac{G}{G_0} = \frac{\triangle OQN}{\triangle OQR} = \frac{MN}{MR} = \cos \phi,$$

or $\quad G = G_0 \cos \phi.$ $\hspace{4cm}$ **6.20**

If we are seeking to understand the vectorial character of the physical quantity 'moment of a force about an axis', then equation **6.20** provides the necessary basis for that understanding. We have derived this equation by considering the moment, about any axis through a fixed point O, of a force of which the magnitude and line of action are completely specified. We find that this moment is greatest (G_0) about an axis ($\overrightarrow{OT}$) which is at right angles to the plane containing O and the line of action of the force; about any other axis through O the moment G is given by $G = G_0 \cos \phi$, ϕ being the angle between that axis and $\overrightarrow{OT}$. This result is the standard result for the resolution of a vector: we conclude, then, that we may regard the moment of a force about any axis as having the characteristics of an axial vector directed along the axis. This conclusion immediately gives vectorial significance to equations **6.18** and **6.19**. The left-hand member of each of these equations is the measure of the moment of a force. The right-hand member is the product of the measures of an angular acceleration and a moment of inertia. We have already concluded that angular acceleration is an axial vector (p. 57); moment of inertia (though the specification of an axis enters into its definition), on the other hand, is obviously a scalar quantity. The equations, then, are, by reference, vectorially homogeneous, as indeed they must be if they are to be physically meaningful.

According to the Newtonian scheme, forces are paired and originate in the mutual interactions of particles. From this point of view, the point of application of a force is of equal importance with its line of action and its magnitude. Suppose, then, that the force **F** of our recent argument has R as its point of application and $\overrightarrow{QR}$ as its line of action as before (see Figure 20). In respect

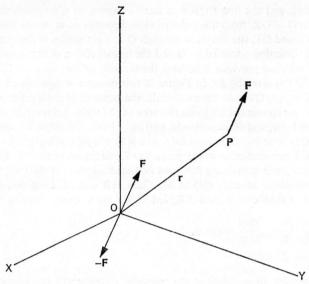

Figure 21

of O as origin, let **r** be the position vector (p. 48) of R. In conventional vectorial notation the moment of **F** about $\overrightarrow{OT}$ is written

$$\mathbf{G_0} = \mathbf{r} \times \mathbf{F}. \qquad \qquad \mathbf{6.21}$$

The right-hand member of equation **6.21** is referred to as the *cross product* or the *vector product* of the two vectors **r** and **F**. It is of the nature of the convention that this product represents an axial vector having direction at right angles to the plane containing the directions of **r** and **F** – and that the positive directions of the three vectors as they appear in order in the equation are related as the positive directions of the coordinate axes are related in a right-handed set. It will be recalled that we encountered a similar formal situation in respect of equation **4.20** giving the acceleration in uniform circular motion (p. 56), and that we left the question of its formulation in vectorial notation for later discussion. We have now provided that discussion. Finally, we note that the 'shorthand' notation of equation **6.21**, providing, as it is claimed to do, information concerning both the magnitude and direction of $\mathbf{G_0}$, must be understood as incorporating our previous result $G_0 = Fd$. If the directions of **r** and **F** include an angle α, then obviously, $d = r \sin \alpha$. Generally, therefore, when a vector product is involved, the magnitude of the product vector is proportional to the sine of the angle between the directions of the two vectors which form the product. On this basis, of course, in the case that we have been considering, the moment of the force **F** about $\overrightarrow{OT}$ is in fact independent of its point of application in the straight line PS.

6.7   Couples and wrenches

In this section, where it is convenient, we shall use the single word *torque* as synonymous with 'moment of a force'. This has the advantage of brevity, and of conformity with much of modern usage.

In the second case that we considered in detail in the last section, that in which, instantaneously, a body was rotating with angular acceleration about an axis through its centre of mass which itself was at rest and unaccelerated, we concluded that the external forces acting on the body exerted finite torque about the axis of rotation, but had zero resultant if transferred, with directions unchanged, to the centre of mass. The simplest force system having these characteristics is a pair of equal and oppositely directed forces of which the lines of action are not collinear. The name *couple* was given to such an associated pair of forces by Poinsot, who was the first to study the properties of such systems in detail, in 1803. For our purposes, the fundamental characteristic of a couple is that the sum of the moments of its constituent forces, about any axis at right angles to the plane containing the lines of action of the forces, is a constant quantity. The measure of this quantity is given by the product of the measure of either force and the measure of the distance of separation of the lines of action of the forces (as can easily be demonstrated). Any couple, then, is represented by a unique direction (perpendicular to the plane just identified) and a unique torque (effective about any axis at right angles to that plane); it may be classified as an unlocalized axial vector, and conventionally symbolized by **G**. In general, in any case in which the centre of mass of a rigid body is, instantaneously, unaccelerated, there is, in respect of the external forces acting on the constituent particles of the body, a unique axial direction of instantaneous torque. The immediate effect of these forces can then be represented in terms of the effects of any pair of equal and oppositely directed forces applied to the body, provided only that the lines of action of the forces lie in a plane at right angles to this direction, and the moment of the couple which they constitute has the appropriate value.

Let us now consider the situation in which the centre of mass is not un-accelerated. If justification is needed for the statements that we have just made, our discussion should provide it; in any event, we should be able to make further assertions of a more general nature for which at present we have no justification. Suppose, then, that an arbitrary point O in a rigid body be taken as origin (Figure 21) – and that, instantaneously, the external force acting on a representative particle of the body situated at P is **F**. Let $\overrightarrow{OP} = \mathbf{r}$. Imagine two equal forces F to be applied at O in directions respectively parallel and anti-parallel to the external force acting at P. This is equivalent to increasing by one the number of pairs of equal and oppositely directed internal forces effective in the body, without in any way altering the over-all cancellation of dynamical effect which is characteristic of the internal-force system as a whole. Suppose that the three forces of magnitude F are now associated as follows: the force **F** at P and the force $-\mathbf{F}$ at O are regarded as forming a couple (the

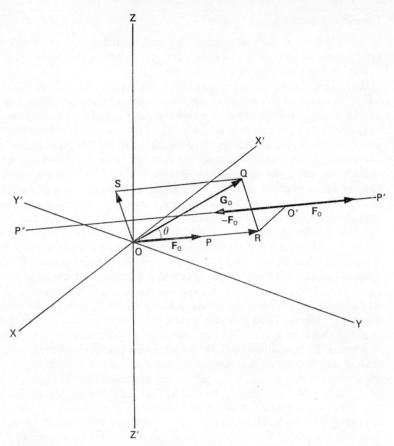

Figure 22

vectorial specification of which is $\mathbf{r} \times \mathbf{F}$); in that case the remaining force $\mathbf{F}$ at O is left unpaired. Suppose, finally, that the external forces acting on all the particles of the body are similarly dealt with. We then have, summing vectorially,

$$\mathbf{F}_0 = \sum \mathbf{F},$$ 6.22

$$\mathbf{G}_0 = \sum \mathbf{r} \times \mathbf{F}.$$ 6.23

Equations 6.22 and 6.23 give the details of a particular 'reduction' of the complete external-force system instantaneously effective. These forces reduce to a force $\mathbf{F}_0$ acting through O, and a single couple of which the axial direction and the moment are specified by $\mathbf{G}_0$. It will be noted that $\mathbf{F}_0$ is independent (in magnitude and direction) of the choice of origin O. In general $\mathbf{G}_0$ is different for different origins. We also note that when $\mathbf{F}_0 = 0$ only the single couple is left: that is the justification, if such were needed, of our statement towards the end of the last paragraph, 'in any case in which the centre of mass of a rigid body is, instantaneously, unaccelerated, there is, in respect of the external forces acting on the constituent particles of the body, a unique axial direction of instantaneous torque.'

Carrying this reduction one stage farther, we represent in Figure 22, with respect to the same origin as before, the 'resultant' force $\mathbf{F}_0$ and the 'resultant' couple $\mathbf{G}_0$ as given by equations 6.22 and 6.23. $\mathbf{F}_0$ is represented, on the force scale of the diagram, by the vector $\overrightarrow{OP}$, and $\mathbf{G}_0$, on the torque scale, by the vector $\overrightarrow{OQ}$. $\angle POQ = \theta$. Let us resolve $\mathbf{G}_0$, in the plane POQ into components of magnitude $G_0 \cos \theta$ and $G_0 \sin \theta$, represented by $\overrightarrow{OR}$ and $\overrightarrow{OS}$, respectively, in the diagram. From R let RO' be drawn at right angles to the plane POQ and of length equal to $(G_0/F_0) \sin \theta$. Through O' the line $P''O'P'$ is now drawn parallel to OP. Finally, let us imagine that, through any point in $P''O'P'$ (in the figure the point O' is chosen), equal and opposite forces F_0 are applied having their lines of action in $P''OP'$. Using essentially the same argument as before (p. 129) we conclude that, so far as their instantaneous effect on the rigid body is concerned, the combination of the force $\mathbf{F}_0$ through O and the couple $\mathbf{G}_0$, is completely equivalent to the combination of an equal force $\mathbf{F}_0$ through any point O' in $P''O'P'$ and a couple of magnitude $G_0 \cos \theta$ having $\overrightarrow{OR}$ as axis. (This is the case because, as we have drawn the figure, the effect of the couple of magnitude $G_0 \sin \theta$ with $\overrightarrow{OS}$ as axis is completely annulled by that of the pair of forces $\mathbf{F}_0$ through O and $- \mathbf{F}_0$ through O'.) We have, then, by our present construction, reduced the single force and the couple of which the axis was inclined to the line of action of the force to a single force of the same magnitude and direction but through another point in the body and a couple of which the axis is parallel to the line of action of the force. Because we started from perfectly general assumptions, we conclude that the external forces instantaneously acting on a rigid body can in all cases be so reduced. This ultimate system, of a force and an associated couple axially concurrent with the force direction, is referred to as a *wrench*. The term was introduced by R.S.Ball (1840–1913) in 1900. In the particular symbolism that we have just been using, the components of the wrench (about the axis $P''O'P'$) are the force of which the measure is F_0 and the couple of measure $G_0 \cos \theta$. In general the linear measure obtained by dividing the measure of the couple by the measure of the associated force is referred to as the *pitch* of the wrench.

As a result of the considerations of the last paragraph we have effected the ultimate in reduction of an arbitrary set of external forces instantaneously acting on a rigid body: in all cases these forces reduce to a wrench. Obviously, the complete specification of the wrench is unique in a given case. For purposes of dynamical calculations, therefore, in general the axis of the wrench cannot be assumed to pass through the centre of mass of the body. Clearly, there will be situations of particular symmetry in which this happens: these situations will be the dynamically 'simple' ones.

6.8   Other derived quantities

In this chapter, assuming our system of kinematical measurement to be already established, we are engaged on the task of basing on that system a system of measurement of significant physical quantities in the field of kinetics (dynamics). Following Newton, we have, so far, introduced and defined one primary non-kinematical quantity, mass, and four derived quantities, density, force, impulse and momentum. Incorporating post-Newtonian developments, we have added two more derived quantities, moment of inertia and torque. In this section, largely for purposes of reference, we give the definitions of certain other such quantities which effectively complete the system. It will be noted that, in making this claim to completeness, we are asserting that a satisfactory system of dynamical measurement can be based on the use of three independent primary units, exclusively. As we have developed the system, these are the units of mass, length and time. This is not the only possible choice, but it is the traditional one – and we shall adhere to it.

We have not, so far (the critical reader may demur) formally defined the standard of mass measurement, as we previously defined the accepted standards of length (p. 28) and time (p. 45). It is appropriate that we should do that now, before proceeding to the other definitions with which we are concerned.

When we were criticizing Newton's Definition 1 (of mass), we wrote (p. 106), 'Assuming mass to be a primary quantity, we should then be at liberty to make an arbitrary choice of standard, assigning unit mass to some body fabricated with particular care and kept for purposes of ultimate reference, and by use of the beam balance and sets of subsidiary standards (or "weights") be in a position to assign a mass-value to any body whatever.' We dismissed this recipe, in the particular context of our argument at the time, for we were concerned then with the logical structure of the system rather than with practical convenience. 'If we do this,' we objected, '... we shall have committed ourselves irrevocably to a particular definition of force.' That definition, however, has now been universally accepted – and as a matter of practical convenience the recipe that we rejected for reasons of logic has been accepted, also, as providing the most accurate means of mass measurement (or, strictly, of mass comparison) that we have been able to devise. The 'body fabricated with particular care and kept for purposes of ultimate reference' is a solid cylinder of platinum–iridium alloy (90% Pt, 10% Ir) of diameter equal to its height (just less than four centimetres). It is kept at the International Bureau of Weights and Measures at Sèvres, near Paris. In 1889 it was designated the International Prototype Kilogramme. Since its designation, the mass of this body has been defined as unit mass (one kilogramme) in each succeeding pronouncement of the General Conference of Weights and Measures (see p. 28). The 'other derived quantities' that we now wish to define are angular momentum, kinetic energy, work and power.

Let us consider a rigid body instantaneously rotating with angular speed ω about a particular axis. Arising from this rotational motion, a representative particle of the body at a perpendicular distance q from the axis of rotation possesses a component of linear velocity $q\omega$, and a component of linear momentum $mq\omega$, in a direction at right angles to both q and the axis of rotation. The moment of this component of linear momentum of the particle about the axis is $mq^2\omega$, and the quantity obtained by summing these moments over all the particles in the body is defined as the instantaneous *angular momentum* of the body about that axis. Denoting the measure of this last quantity (which is evidently an axial vector) by the symbol H, we have

$$H = \sum mq^2\omega = I\omega, \qquad\qquad 6.24$$

I being the moment of inertia of the body about the axis in question. For obvious reasons, the angular momentum is sometimes referred to as the moment of momentum of a body about an axis: more cogently the alternative term is traditionally used when a system of freely moving particles, rather than a rigid body, is involved.

The term *kinetic energy* was introduced into physics as recently as 1862, by William Thomson (Lord Kelvin) and P. G. Tait (see p. 50) in a joint article in *Good Words*, but the physical quantity which was so designated had been recognized as significant in many situations since the time of Leibniz and Huygens. Newton, writing in Latin, had used the term *vis viva* – and the term had been taken over consistently by later writers, without translation, until Thomas Young (1773–1829) suggested 'energy' as an English equivalent in 1807. Young's suggestion, however, proved unacceptable to many: it was not until it was revived (with the necessary adjectival qualification), more than half a century later, that it was finally adopted.

Fundamentally, we define kinetic energy in relation to a single particle of mass m, instantaneously moving with speed v. We say that the measure of its kinetic energy is then $\frac{1}{2}mv^2$ (Newton, in fact, said that the measure of its *vis viva* was mv^2, but this difference in definition is insignificant). On the basis of our present definition, kinetic energy is obviously a scalar quantity. Also, it is easy to extend it to a system of particles and to show that the total kinetic energy of such a system is uniquely divisible into two parts – the energy of the relative motion of the particles with respect to their centre of mass, and the kinetic energy which a composite particle would have if it contained the mass of all the particles and moved with the speed of the centre of mass. Taking a rigid body as a special case of a system of particles, as we have done on previous occasions, we conclude from the last statement that if, instantaneously, the centre of mass of a rigid body is moving with speed v, and the body is rotating about an axis through the centre of mass with angular speed ω, the total kinetic energy of the body is $\frac{1}{2}Mv^2 + \frac{1}{2}I\omega^2$, M being the mass of the body and I the moment of inertia about the relevant axis. (The second term in this last expression follows from the fact that, in our previous notation, the contribution to this term in respect of a representative particle of the body is $\frac{1}{2}mq^2\omega^2$.)

Qualitatively, *work* is said to be done when resistance is overcome. This is too much like a statement of everyday experience to serve as a satisfactory definition of a physical concept, though it is a true statement when given the necessary precision. We say, rather, that when the point of application of a force **F** moves through a distance s work is done 'by the force' in amount $Fs \cos \theta$, θ being the angle between the vectorial directions of **F** and **s**.

We have just given the standard definition of work as a physical quantity, but we have to admit that, on any fundamental Newtonian assessment, the phrase 'the point of application of a force' is peculiarly evasive. Forces occur exclusively as pairs of equal and oppositely directed actions between pairs of particles. At the fundamental level, the movement of the point of application of a force is simply the movement of a particle – a free particle, or a constituent particle of a gross body. Let us examine, therefore, the simplest possible case of the uniformly accelerated rectilinear motion of a particle of mass m. In the standard kinematical notation, equation 4.7 relates the velocities of the particle at the beginning and end of a linear interval s:

$$v^2 = v_0^2 + 2as.$$

Introducing the measure of the mass into the equation, we have

$$\frac{1}{2}mv^2 - \frac{1}{2}mv_0^2 = mas. \qquad \text{6.25}$$

In equation 6.25, according to our last definition, the left-hand member represents the increase of kinetic energy of the particle; according to the present definition, the right-hand member represents the work done by the force which is responsible for the acceleration a of the particle in its rectilinear motion. This result is sufficient in itself to show that, in any self-consistent system of measurement, work and kinetic energy must be measured in the same units; in the particular case to which it refers it also shows 'what becomes of' the work which is done. In this case work is expended in increasing the kinetic energy of the particle on which the force acts. A similar statement cannot validly be made in every case, but that it can be made in any is sufficient for our immediate purpose.

Work, being measurable in the same units as kinetic energy, must be a scalar quantity, as the other is. We note, however, that the measures of two vector quantities enter into its specification – and that the relative directions of these vectors are also involved. In vectorial notation, this situation introduces us to a new type of vector product (see p. 128). We write, in respect of the work W,

$$W = \mathbf{F}.\mathbf{s}, \qquad \text{6.26}$$

using the dot (rather than the multiplication sign) as indicating the product-forming operation. In relation to equation **6.26**, W is said to be the *scalar product* of the two vectors **F** and **s**. As we have previously stated, when numerical measures only are in question, $W = Fs \cos \theta$, θ being the angle between the directions of the vectors concerned.

Power is defined very simply: it is the instantaneous rate of performance of work. It is, of course, a concept having technological connotation, but the physical quantity which derives from it is of considerable use in the quantitative treatment of many situations of fundamental interest to physicists also.

We made the claim that in this section we should complete the definitions of those physical quantities which are necessary (and essentially sufficient) for the formal development of the dynamics of particles and rigid bodies. It will be convenient at this point, then, to give a list of the primary and derived quantities that we have dealt with, indicating the units in which they are measured in the M.K.S. (S.I.) system.

Length	metre	m	Angular momentum	—	$\mathrm{kg\ m^2\ s^{-1}}$
Time	second	s	Force	newton	$\mathrm{N\ (kg\ m\ s^{-2})}$
Mass	kilogramme	kg	Torque	—	$\mathrm{kg\ m^2\ s^{-2}}$
Density	—	$\mathrm{kg\ m^{-3}}$	⎧Kinetic energy	joule	$\mathrm{J\ (kg\ m^2\ s^{-2})}$
Moment of inertia	—	$\mathrm{kg\ m^2}$	⎩Work		
⎧Momentum	—	$\mathrm{kg\ m\ s^{-1}}$			
⎩Impulse			Power	watt	$\mathrm{W\ (kg\ m^2\ s^{-3})}$

It will be noted that, in the list that we have given, it appears at first sight that the units of measurement of torque and kinetic energy are the same ($\mathrm{kg\ m^2\ s^{-2}}$), though a distinctive name (joule) is given only to the unit of kinetic energy (or work). This apparent paradox arises from the fact (which we have already established – see pp. 127, 133) that the vectorial characteristics of these two quantities are different: torque is an axial vector, whereas kinetic energy is a scalar quantity. There can be no physical situation in relation to which it is significant to add together the measures of a torque and a kinetic energy characteristic of the system involved. If it is desired to express in words the magnitude of a particular torque, the unit in question may conveniently be referred to as the 'newton metre'.

6.9 Conservation laws

At least since the time of Galileo – though with no secure justification in logic – natural philosophers have frequently invoked the criterion of simplicity in the formulation of natural law. They have professed the belief that, if only the 'correct' concepts can be developed, the processes of the physical world will be found to be susceptible of precise quantitative description in terms of simple algebraic equations relating the measures of the physical quantities so defined (see p. 51). It is one of the remarkable features of the Newtonian scheme that it is characterized by a basic simplicity of a very fundamental kind – indeed, this feature may be seen as the foundation of its 'success'. However that may be, we examine in this section the 'conservation laws' which expose the nature of this feature of simplicity in its most obvious form: the concepts of mass and (linear) momentum which Newton introduced are such that the total amount of mass, in any isolated system remains constant throughout time – and, similarly, the vector sum of the momentum remains constant. These concepts, in fact, are of such a character that, in their global reference, they identify quantitative aspects of permanence in the real world (see p. 18).

The principle of the conservation of mass is implicit in the statement of Law 1 of our revised formulation (p. 111): 'It is possible to assign to every identifiable portion of matter in the universe a unique mass-measure....' In the later discussion of section 6.5 we had to admit that this statement was possibly over-optimistic at the time that it was made, but we were able to justify it in the end, demonstrating the additivity of mass on the Newtonian scheme, once we had established an operationally based definition of this quantity in respect of a composite body. Within the framework of Newtonian mechanics there is nothing further to be said regarding this fundamental law of nature, except to utter the caution that its relevance to reality presupposes the permanence as identifiable entities (see Definition 1, p. 111) of the 'primordial particles' which are the ultimate constituents of material bodies. For this reason the principle of the conservation of mass has sometimes been referred to as the principle of the indestructibility of matter.

The principle of the conservation of linear momentum is effectively contained in the statement of Law 2 of our new formulation – at least when the concepts of mass and force have already been defined, and the law takes on the alternative form that we gave it in later discussion (p. 112): 'When N particles form an isolated system ..., there is ... a unique way of resolving the total forces acting on the particles ... so as to produce $\frac{1}{2}N(N-1)$ pairs of oppositely directed component forces ... for each of [which] pairs the associated component forces are equal in magnitude.' Let us consider any one of the $N(N-1)$ component forces referred to in this statement of the law. If m_r is the mass of the particle on which it acts, this force is instantaneously generating linear momentum, at a rate equal to its own measure, in its own direction (in general, for one-dimensional motion, $F \equiv ma \equiv m\dot{v} \equiv \dot{p}$). The law states that there must be another component force (acting, say, on the particle of mass m_s) of precisely the same magnitude, oppositely directed. Obviously, these two forces are instantaneously generating linear momentum at the same rate, but in opposite directions. A similar result holds for each of the other pairs of component forces effective at the instant in question, and we finally conclude that the over-all rate of generation of linear momentum in the isolated system is zero: the vector sum of the linear momenta of the individual particles of the system remains constant throughout time.

An analogous result holds for the total angular momentum (or the vector sum of the moments of momentum of the individual particles) of an isolated system, whatever axis is chosen, once we have accepted Law 3 (p. 123): 'The equal and oppositely directed forces of interaction of any pair of particles are effective in the same straight line.' Without this additional law, the principle of the conservation of angular momentum in an isolated system would have no formal basis; once the law has been accepted, the principle stands established – and it need not detain us further at this stage.

Having reviewed the position in respect of the quantities mass, momentum and angular momentum, it is natural that we should inquire concerning the status of the fourth quantity which represents a dynamical attribute of a body in motion, namely its kinetic energy. The simple question is: Is the principle of conservation of kinetic energy valid for an isolated system? The direct answer is that it is not. If we consider an isolated system of two particles which interact with a force of mutual attraction at all distances, clearly the total kinetic energy of the system continually increases during any period in which the particles are approaching one another, and decreases during any period in which their separation is increasing. Again, we have the phenomenon of two-body collisions, which was examined by Newton (see section 5.4). Using the notation of our previous discussion, which referred to the case in which one body, of mass m_2, was initially at rest, and the collision was 'direct' (rather than 'oblique'), we have, for that case,

$$m_2 v_2' = m_1(v_1 - v_1'), \qquad\qquad 6.27$$

$$v_2' - v_1' = ev_1. \qquad\qquad 6.28$$

Equation 6.27 simply repeats equation 5.4; equation 6.28, on the other hand, represents symbolically what previously was expressed only in words – the coefficient of restitution (p. 95), denoted by e, is a constant for the system (and numerically is never greater than unity). Incidentally, we note, as is appropriate in the present context, that equation 6.27 essentially represents the law of conservation of linear momentum as applied to the collision. In respect of kinetic energy, solution of equations 6.27 and 6.28 for v_1' and v_2' finally gives

$$\tfrac{1}{2}m_1v_1'^2 + \tfrac{1}{2}m_2v_2'^2 = \tfrac{1}{2}m_1v_1^2 - \frac{1}{2}\frac{m_1 m_2}{m_1 + m_2} v_1^2(1 - e^2). \qquad\qquad 6.29$$

Equation 6.29 shows that, except when $e = 1$, the sum of the kinetic energies of the two bodies after collision is less than the kinetic energy of the moving body before collision. Over-all, then, kinetic energy is not conserved, for gross bodies, in collision processes – except when $e = 1$.

The special case $e = 1$, though no two solid materials precisely match its formal requirement, is not so trivial as to be unworthy of further consideration. Newton himself was sufficiently perspicacious to attach importance to it as an ideal to which gross bodies never completely attain but which the ultimate particles of matter may be assumed to exhibit. He wrote, 'Bodies absolutely hard return from one another with the same velocity with which they meet' – and in another place, 'It seems probable to me ... that these primitive Particles being Solids, are incomparably harder than any porous Bodies compounded of them, even so very hard, as never to wear or break in pieces ...' (Opticks, 2nd edition, 1717, query 31). Newton, then, imagined the ultimate particles of matter to be such that in collisions between them kinetic energy would be conserved. Nearly a century and a half later the simple kinetic theory of gases was successfully based on this fundamental assumption.

In the collisions of gross bodies, during the period of physical contact, the bodies are instantaneously deformed (see pp. 94, 99). During this period their relative velocity is rapidly reduced to zero and is later restored to its final value (its direction having been changed). By definition (p. 95), the coefficient of restitution measures the extent of this restoration. In the ideal case the restoration is complete. During the period of contact, therefore, kinetic energy is temporarily lost, only to be restored again – in the ideal case completely. In these observations we recognize the beginnings of views concerning the transformation of 'energy' from one form to another – and the indication, possibly, of a law of conservation of energy, if all the forms which are mutually convertible, one into another, can be satisfactorily identified. Hermann L. F. von Helmholtz (1821–94) was the first to take the bold step of proposing such a law of universal conservation in 1847 – when all the possible forms certainly had not been identified (nor a full understanding been achieved in relation to those that had). Time has proved the soundness of von Helmholtz's intuition in this matter, but we cannot pursue the matter farther here. We restrict ourselves, instead, to one further remark on the collision phenomenon through which we introduced these considerations. It is natural to assume (and, indeed, possible to construct a detailed theory to justify that assumption) that kinetic energy is transformed into 'energy of deformation', in the initial phase of the contact of the colliding bodies, to be transformed back again into kinetic energy as the deformation decreases in the final phase. When the body is deformed, the configuration of its constituent particles changes in the region of deformation (p. 117): the 'stored' energy, then, is configuration-dependent energy. In 1853, W. J. M. Rankine (1820–72) introduced the term *potential energy* to denote that portion of the energy of any system which is configuration dependent – that is, determined by the relative positions of the particles of the system.

6.10   **Newtonian particles and real atoms**

Newtonian dynamics is based on a simple model of reality – on ideal rigid bodies constituted of primitive particles which, whilst having finite extension (p. 111), are yet considered incapable of rotatory motion (p. 19). Today, our model of reality is very different. The twentieth-century physicists' atom (the counterpart of Newton's particle) is almost every bit as 'real' as the gross bodies of Newton's philosophy: it is a complicated structure, itself constituted of protons and neutrons and electrons, and in general it is certainly not incapable of rotation. We wonder, perhaps, how so crude a model as Newton's could have been so successful – particularly since fundamentally (as we have tried to stress in our exposition of it) Newton's dynamical system involves positive assertions concerning the nature of the interactions between the primitive particles themselves. In this connexion we must not lose sight of the fact that the initial elucidation of the principles of electrostatics, and the later discoveries of the electron and the proton, the 'elementary carriers' of electric charge – and of the neutron, too – were themselves based on Newton's laws.

With these considerations in mind, we ask the question whether it is possible that the laws of motion – particularly in the form in which principles of conservation emerge – are almost model independent. Possibly it is sufficient that our model should provide that the forces ultimately originate in entities smaller by very many orders of magnitude than the 'instruments' by which we observe their effects: if this view is accepted, the precise nature of the entities may be largely irrelevant. It may well be that there is significance in this suggestion.

When we introduced Law 3 (p. 123) – and so effectively reconciled observation and theory in rigid-body dynamics – at the same time establishing the principle of conservation of angular momentum in the Newtonian system (p. 137), we did not go the whole way with the followers of Newton. Adopting the traditional view of the Newtonian particle as incapable of rotatory motion, Newton's followers have, in effect, consistently worded this statement as follows: 'The equal and oppositely directed forces of interaction of any pair of particles are effective in the straight line joining the particles.' If pressed with the observation that Newtonian particles must in the end be assumed to have finite extension, these writers would probably concede that what is in question in the law is the straight line joining the centres of mass of the particles. We made no such specific assertion: advisedly we wrote, instead, 'in the same straight line'. That was sufficient for our immediate purposes, when only the internal forces acting between the constituent particles of a rigid body were concerned. In failing to take the last step with Newton's disciples, we left the way open intentionally, for a final amendment of our law. We quote the amended form, forthwith, and then proceed to a brief discussion.

Law 3 (amended). In addition to the force pair identified on the basis of Law 2 and Definitions 3 and 4, the total interaction of any pair of particles in general includes a pair of couples, also, in such a way that the total interaction is a wrench pair in which the two wrenches (one acting on each particle from the other) are of the same pitch and cheirality and involve equal forces oppositely directed along a common axis. [We note, for ease of reference, that the term *pitch*, in relation to a wrench, was defined on p. 131 and that *cheirality* was defined on p. 24.]

The obvious comment on our amended law is that it leaves open the possibility of rotatory motion for the ultimate particles of our dynamical model. We have, in effect, so modified Newton's original model that it matches experience somewhat more closely than before. But in framing the amended version of the law we have been careful not to jettison the conservation principles. The constituent forces of the wrench pairs still cancel completely in their dynamical effects within an isolated system – and the newly introduced couples generate angular momentum at a net rate that is zero when the appropriate vector sum is taken over the whole system concerned. We do not claim that the amended law describes a state of affairs which, macroscopically considered in the context of the interaction of gross bodies, is essentially different from that described in the traditional form of the law (p. 139) that we have ascribed to the 'followers of Newton', but we have at least satisfied our own conscience concerning the status of the ultimate particles of matter in our conceptual scheme. In reality, these entities have a diversity of properties greater than was envisaged by Newton. We must leave it at that.

Further reading

A. B. Arons, *Development of the Concepts of Physics* (Chapters 1–21), Addison-Wesley, 1965.

N. Feather, 'Additivity of mass in Newtonian mechanics', *American Journal of Physics*, vol. 34 (1966), pp. 511–16.

M. Jammer, *Concepts of Force*, Harvard University Press, 1957.

M. Jammer, *Concepts of Mass*, Harvard University Press, 1961.

R. B. Lindsay, *Physical Mechanics* (3rd edn, Chapters 1–8), Van Nostrand, 1961.

R. B. Lindsay and H. Margenau, *Foundations of Physics*, Wiley, 1936.

E. Mach, *The Science of Mechanics* (1883) (6th edn), Open Court, 1960.

Units and Standards of Measurement employed at the National Physical Laboratory, I, H.M.S.O. (revised periodically).

Chapter 7
Gravitation

7.1 Terrestrial gravity and universal gravitation

We have given some account already of the early experiments on the motion of bodies 'under gravity' near the surface of the earth (section 5.3), and of the way in which Newton was able to bring the results of these experiments, and those of the experiments on collisions (section 5.4), within a single scheme of interpretation (section 6.1). Precision was thereby given to the concepts of force and mass, and the bold assertion was made, in the form of a law of nature, that forces in the real world occur only as pairs of equal and oppositely directed members. So, also, for the first time, it came to be appreciated that the weight of a body – the force of gravity acting on it – is proportional to its mass (for the accelerations of all bodies are the same, at the same place (equation **6.6**)).

Yet, if experiments on falling bodies contributed to the ultimate formulation of the laws of motion, they did not provide the only evidence concerning the nature of gravitational phenomena which Newton took into account in fashioning those laws. As we have already remarked (p. 102), Newton was as much concerned, at the time, 'with the planets in their orbits as with stones falling to the ground, or pendulums oscillating in his laboratory'. The third law of motion – the fundamental law which we quoted in paraphrase above – has no merely terrestrial reference: evidence from astronomical observation contributed to the aspect of complete generality which it finally assumed (p. 103). Newton's third law of motion is to be found in Book 1 of *Principia*. Book 3 is written around the law of universal gravitation, an exemplar of the other: 'Between each pair of particles in the universe there exists a force of gravitational attraction, in magnitude directly proportional to the product of the masses of the particles and inversely proportional to the square of their distance of separation.' So far as the international community of men of science was concerned, the two laws were published together, in July 1687.

We shall in due course have to consider the totality of evidence on which Newton based the law of universal gravitation, but in this introductory section there are certain matters that may be disposed of before we proceed farther. It may be useful to give a very brief conspectus of the attitude of pre-Newtonian philosophers to the group of phenomena that we now regard as *gravitational*, it may be instructive to expand our statement that the law of universal gravitation is 'an exemplar' of the third law of motion, and it will certainly be convenient if we deal with the problem of the *centre of gravity* (p. 114) – which arises directly from the law of gravitation as we have stated it – at this early stage.

It is a fact of history that the first speculative formulation of a principle of universal gravitation was made in 1643, a few months after the birth of Newton, by Gilles Personne de Roberval (1602–75), professor of mathematics at the Royal College of France in Paris. Roberval canvassed the view that every particle of matter in the universe in fact attracts every other such particle. However, there was nothing quantitative about his suggestion, and in consequence it was basically ineffectual. Yet, in making the suggestion at all, Robreval shares with Descartes (whose *Principles of Philosophy* was published in the following year – see p. 92) the credit of resolutely abandoning the traditional cosmological view. Until the time of Roberval and Descartes, the Aristotelian dichotomy – the distinction between the laws of the sub-lunar and the celestial worlds – was accepted without question. Until that dichotomy was healed, the concept of universal gravitation could not emerge; as long as it was maintained, natural philosophers were restricted in their speculations to the making of suggestions regarding the motion of the moon (the motions of the other celestial bodies were 'perfect' motions). In fact, the general notion of sub-lunar attractions had been current at least from the first half of the thirteenth century in Europe, arising no doubt from growing familiarity with the properties of the lodestone and the artificial magnets that the mariners used. Thereafter, until the time of William Gilbert (1544–1603), when philosophers discussed the possible attraction of the earth for the moon – or of the moon for the ocean waters – they were apt, quite innocently of quantitative considerations, to suggest 'magnetism' as the cause. It remained for Roberval – and more particularly for Newton – to disabuse them of such fanciful notions.

Our second present concern is with the statement that the law of universal gravitation is an exemplar of Newton's third law of motion. By this we imply that the 'laws of motion' collectively provide the means of identifying the forces which are effective in the physical world – and the third law, in particular, makes the specific assertion that these forces always occur in equal and oppositely directed pairs. We further imply that the law of universal gravitation constitutes a special case in which this assertion is exemplified, indeed the first such case to be recognized in its generality. It states, in effect, that there are forces (which henceforth will be denoted as *gravitational*), which are effective as between all pairs of particles in all circumstances, the magnitudes of which are directly related to the masses of the particles concerned. The forces of gravitation, according to the law, issue from the 'massiness' of matter.

The history of fundamental physics in the two centuries following the publication of *Principia* is largely the history of the processes of recognition of other types of force, the magnitudes of such forces being related to other physically identifiable attributes of particles – less indwelling attributes, perhaps, than the attribute of mass. Thus, small particles of matter, in special circumstances, are found to exhibit forces of attraction (and repulsion) which are enormously greater than the gravitational forces which are effective between them. Ultimately, a whole group of phenomena comes to be recognized as justifying the special designation *electrical*. Finally, the evidence is codified in Coulomb's law (1785): 'Between any pair of particles charged with electricity there exists a force of attraction (or repulsion), in magnitude directly proportional to the product of the quantities of electricity associated with the particles and inversely proportional to the square of their distance of separation.' Later (1832), there is a similar law in respect of *magnetic* forces – and so on.

Even the casual reader cannot but notice the essential similarity of form of the laws of Newton and Coulomb, as we have expressed them here. There is the obvious (though basically trivial) difference in the opening phrases – 'Between each pair of particles in the universe' and 'Between any pair of particles charged with electricity', which reflects the fact that all particles are not 'charged with electricity' all the time, whereas every particle has its mass as a permanent attribute – but, apart from that, the two statements are structurally identical. It is the more important, therefore, that we should emphasize the subtle difference which informs these statements. The law of universal gravitation states a quantitative relation between the measures of physical quantities of three kinds (force, mass, distance) all of which relate to concepts already defined. In respect of Coulomb's law, on the other hand, physical quantities of three kinds (force, quantity of electricity, distance) again being involved, one of them relates to a concept (electric charge) which has not previously been defined. (In asserting that the laws state 'quantitative relations', we are for the moment disregarding any 'multiplying constant' that may be in question, in one or both of them.)

It is a fact of history, as every elementary textbook of electrostatics of the nineteenth century would insist, that the definition of the unit of charge (quantity of electricity) was based on Coulomb's law in the first instance: 'Unit charge is that charge which at unit distance from a precisely similar charge (in vacuum) repels it with unit force.' Later, under the criticism of use, the definition ceased to be regarded as wholly satisfactory: it implies, unacceptably, that 'quantity of electricity' – assuredly a non-dynamical magnitude – can be characterized completely in dynamical terms. However, that aspect of the matter is not our concern here (it is bound up with the question of 'multiplying constants' to which we have already referred). What we are concerned with, rather, is the possibility that an 'alternative' definition of mass may be based on Newton's law of universal gravitation, in the same way as the definition of charge was originally based on Coulomb's law.

Let it be said, at once, that any definition of mass (or quantity of matter) which we may devise in this way cannot replace the definition in terms of the inverse ratio of the mutual accelerations of particles to which we are already committed. That definition is necessary for our definition of force – and we must know what we mean by force before the law of universal gravitation itself has any meaning. Originally, when we were establishing our primary definition, we were careful to use the term *inertial mass*; later in the discussion we abandoned the adjective for sake of brevity. We must reintroduce it here, otherwise our argument will be confused. We have, then, a perfectly clear definition of inertial mass; the question is whether there is another physical quantity, fundamentally different in character from inertial mass, but intimately related to it, which may be defined in terms of the law of gravitational action. Let us define the *gravitational mass* of a particle, of measure μ, by the equation

$$F = \frac{\mu_1 \mu_2}{r^2}.$$ 7.1

Here we are disregarding the possible inclusion of a 'multiplying constant', as Coulomb did, and we are saying that the measure of the gravitational force acting between two particles is given precisely by the product of the measures of the gravitational masses of the particles divided by the square of the measure of their distance of separation (see p. 143). Effectively, we are defining the unit of gravitational mass by this statement: 'Unit gravitational mass is such that two particles each of this mass attract one another with unit force at unit distance of separation.' This is an admissible definition – provided that the problem of the 'multiplying constant' is ignored.

Let us now consider the problem of weight from this point of view. Obviously, the weight W of a particle of gravitational mass μ, at any point on the earth's surface, is given by

$$W = \mu \Gamma,$$ 7.2

where Γ stands for the vector sum of contributions, each in its appropriate direction, of quantities having measures such as μ'/r^2, μ' being the gravitational mass of a particle of the earth's substance, and r the distance of that particle from the 'experimental' particle that we are considering. If g is the acceleration due to gravity, measured in respect of our experimental particle at the location in question,

$$\mu \Gamma = mg,$$

m being the measure of the inertial mass of the particle. Clearly, the quantity Γ is characteristic of the point on the earth's surface where the measurements are being carried out. If we take it as a fact of experience that the acceleration due to gravity is also characteristic of location, being the same for all bodies (within the limits of experimental uncertainty) at any arbitrary location, then there must surely be a constant ratio as between the measures of the gravitational and inertial masses of bodies generally. That this appears to be so, we accept as an empirical natural law. Much thought has gone into attempts to understand how this result comes about, but we cannot follow the matter farther here. Suffice to say that the comparison of weights, according to our present discussion, provides no unequivocal information except regarding the ratio of gravitational masses. This is the basis of '[the other], more subtle, point of criticism' that we noted for later elaboration when we rejected the possibility (p. 106) that a satisfactory definition of (inertial) mass could be based on Newton's explanatory gloss, '[The mass] is known for each body by means of its weight.'

Our third concern in this section is with the problem of the *centre of gravity*. The proposition for discussion is that 'for any body (which is not too large) near the surface of the earth the whole weight of the body reduces to a single force acting through a unique point fixed with respect to the body, whatever the orientation of the body.' As we have already stated, this proposition was first established with formal rigour by John Wallis in 1671 – sixteen years before the publication of Newton's *Principia*. We shall not, of course, assume a similar ignorance of Newtonian mechanics in discussing it now.

Let us first, therefore, examine the proposition in a dynamical context. We accept our previous results in relation to the general motion of a rigid body: first, that the centre of (inertial) mass of the body moves as if all the mass of the body were concentrated there and all the external forces applied there in their actual directions – and second, that the instantaneous accelera- tion of angular motion of the body depends on the sums of the moments of the external forces about the principal axes of inertia through the centre of mass of the body (and is identically zero when these torques are zero). Now let us imagine a rigid body released from rest (in any initial orientation) near the surface of the earth. Each particle of the body, if free, would have the same linear acceleration (assuming a constant ratio as between the measures of inertial and gravitational masses for all the particles). When the body is at rest (and we are assuming that the same is true when the body is in accelerated linear motion) the internal forces cancel completely on each particle. There- fore the particles, in their state of mutual coherence in the rigid body, move from rest, all with the same linear acceleration, in the situation that we have described. There is no rotational acceleration, therefore, and consequently there can be no resultant torque. In free fall, the gravitational forces acting on a 'small' body near the surface of the earth effectively reduce to a single force acting through the centre of inertial mass of the body. This purely verbal argument, it will be noted, identifies the centre of gravity with the centre of inertial mass of a body, it being assumed that the measures of inertial and gravitational mass are in constant ratio.

We now consider the problem in a statical context, using the principle of moments (of much greater antiquity than Newton's *Principia*, as we have already indicated, see p. 123) – and Cartesian rather than vector notation (as, effectively, Wallis was compelled to do). Let us imagine a rigid body having one point fixed in space at O (Figure 23) near the surface of the earth. (We are permitted, in a 'thought experiment', to imagine a vain thing!) Suppose that O X, O Y, O Z are rectangular axes through O, also fixed in space, O Z being drawn vertically upwards – that is in the direction opposite to the common

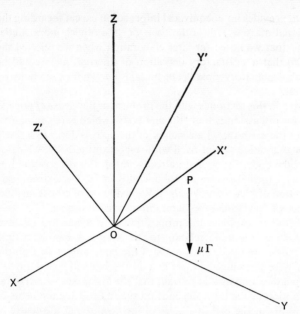

Figure 23

direction of all the forces of gravity acting on the particles of the body. Let OX', OY', OZ' be rectangular axes fixed with respect to the body, and, when the body has come into a position of equilibrium under gravity, let the directions of these 'body axes' (in relation to the 'space axes') be specified by direction cosines α_1, β_1, γ_1, α_2, β_2, γ_2 and α_3, β_3, γ_3, respectively. Then, if the coordinates of a representative particle of the body, of gravitational mass μ, are (x', y', z') with respect to the 'body axes', and (x, y, z) with respect to the 'space axes', we have

$$\left. \begin{array}{l} x = \alpha_1 x' + \alpha_2 y' + \alpha_3 z', \\ y = \beta_1 x' + \beta_2 y' + \beta_3 z', \\ z = \gamma_1 x' + \gamma_2 y' + \gamma_3 z'. \end{array} \right\} \qquad \textbf{7.3}$$

Now, the measure of the gravitational force acting on the representative particle (at P, Figure 23) is $\mu\Gamma$ in our present notation (see equation 7.2), and its direction is parallel to $\overrightarrow{ZO}$. Identically, then, for the whole body, the total torque about OZ is zero. If the body is in equilibrium, the total torque must also be zero about both OX and OY. These two conditions imply

$$\sum \mu\Gamma y = \sum \mu\Gamma x = 0. \qquad \textbf{7.4}$$

Because Γ is characteristic of location only, from equations 7.3 and 7.4 we obtain

$$\begin{array}{l} \alpha_1 \sum \mu x' + \alpha_2 \sum \mu y' + \alpha_3 \sum \mu z' = 0, \\ \beta_1 \sum \mu x' + \beta_2 \sum \mu y' + \beta_3 \sum \mu z' = 0. \end{array} \qquad \textbf{7.5}$$

These equations have a particular as well as a general solution. The particular solution is given by

$$\sum \mu x' = \sum \mu y' = \sum \mu z' = 0. \qquad \textbf{7.6}$$

This solution, which does not involve the direction cosines of the body axes, OX', OY', OZ', obviously implies that equilibrium is possible for all orientations of the body if O, the single point of support, is suitably chosen with respect to the body. Equations **7.6** provide the recipe for the location of that point. Clearly, that a point having these particular properties can be found implies that all the external (gravitational) forces acting on the body effectively act through the point.

In order to expose the general solution of equations **7.5**, we must necessarily suppose that O, the point of support, is not the particular point in the body through which the resultant of all the external forces act. Under these circumstances, let us write

$$\left. \begin{array}{l} \sum \mu x' = \bar{x}' \sum \mu, \\ \sum \mu y' = \bar{y}' \sum \mu, \\ \sum \mu z' = \bar{z}' \sum \mu, \end{array} \right\} \qquad \textbf{7.7}$$

then the general solution that we are looking for is given by the equations

$$\begin{array}{l} \alpha_1 \, \bar{x}' + \alpha_2 \, \bar{y}' + \alpha_3 \, \bar{z}' = 0, \\ \beta_1 \, \bar{x}' + \beta_2 \, \bar{y}' + \beta_3 \, \bar{z}' = 0. \end{array} \qquad \textbf{7.8}$$

Comparison of equations **7.8** and equations **7.3** shows that, if G is the point in the body whose coordinates $(\bar{x}', \bar{y}', \bar{z}')$ with respect to the body axes are given by equations **7.7**, then when the body is in equilibrium suspended from its chosen point of support, $(\bar{x}, \bar{y}, \bar{z})$ the coordinates of G with respect to the space axes are such that $\bar{x} = \bar{y} = 0$. In other words, the point G lies in OZ (Figure 23) – either vertically above, or vertically below, the point of support O. If all the external (gravitational) forces acting on the body effectively act downwards through G, we recognize the former possibility as representing a situation of unstable equilibrium, the latter possibility as representing a situation of stable equilibrium.

In equations **7.7**, on the basis of statical considerations, we have now derived expressions for the coordinates of the *centre of gravity* of a rigid body in terms of the spatial distribution of gravitational mass within the body. In form, these equations are identical with the equations specifying the location of the *centre of mass* in terms of the spatial distribution of inertial mass (pp. 114, 119). Furthermore, because of the empirical constancy (for any location) of the ratio of the measures of gravitational and inertial mass, we may conclude that in all practical situations the centre of mass and the centre of gravity of a body coincide.

147   Terrestrial Gravity and Universal Gravitation

7.2   Universal gravitation: the astronomical evidence

We have quoted Maxwell's definition of a material particle (p. 19), with its addendum, 'Thus in certain astronomical investigations the planets, and even the sun, may be regarded each as a material particle....' It is precisely by so regarding these bodies (as 'point masses') that Tycho Brahe and Kepler provided Newton with the bulk of the astronomical evidence which he used in the process of formulating the law of universal gravitation. In the end, as we shall see (section 7.3), Newton tested his law numerically in relation to the 'earth–apple–moon' system – in which certainly all the bodies could not be regarded merely as point masses – but the most clear-cut evidence that he was able to cite in its favour remained that which related to the larger system of the sun and the planets.

In an earlier chapter (p. 40) we referred to the three 'laws of planetary motion' into which Kepler finally (in 1619) reduced the systematic observations of Brahe. We did not then quote the laws *verbatim*. Now, when their detailed content is our particular concern, we do so forthwith, using modern phraseology:

Law 1. Each planet moves in an ellipse with the sun at one focus.

Law 2. For each planet the line from the sun to the planet sweeps out equal areas in equal times.

Law 3. The squares of the periodic times of the planets in their orbits are as the cubes of their mean distances from the sun.

It will be noted, at the outset, that Kepler's three laws are statements of empirical regularities observed to hold amongst the measures of purely kinematical quantities – and that the statements are made largely in geometrical terms. However, a law geometrically expressed is not in the most convenient form for dynamical interpretation, so the first aim of the physicist must be to translate the laws into other terms which are more directly significant for his purposes. In particular, if the laws can be exhibited as specifying how the accelerations of the planets in their orbits vary with their distances from the sun, then he will be able to proceed immediately to their dynamical evaluation. As Newton eventually showed, the most straightforward translation into terms of acceleration is possible with Law 2.

In Figure 24 let A, B and C represent three points in a small element of the orbit of a planet about the sun (S in the figure). Let it be supposed that the planet passes from A to B and from B to C in equal times. Imagine A B produced to C', so that $BC' = AB$. Then, as to area,

$$\triangle SBC = \triangle SAB \quad \text{(Law 2)},$$

and $\triangle SBC' = \triangle SAB$ (by construction).

In consequence, $\triangle SBC' = \triangle SBC$,

and it may be concluded that C'C is parallel to BS. So far, we have been

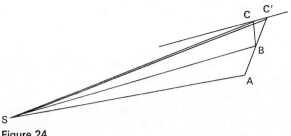

Figure 24

arguing entirely geometrically, but because we have stipulated that equal times are involved in the passage of the planet from A to B and from B to C, we may now regard $\overrightarrow{AB}$ and $\overrightarrow{BC}$ as vector representations of the mean velocities of the planet in these successive intervals of time. Vectorially, also, $\overrightarrow{BC'} = \overrightarrow{AB}$. Thus $\triangle BC'C$ may be regarded as the vector triangle giving the increment of average velocity of the planet ($\overrightarrow{C'C}$) as between the first interval (when the velocity is $\overrightarrow{BC'}$) and the second (when it is $\overrightarrow{BC}$). In the limit, therefore, when the time length of each interval is reduced to zero, the direction of $\overrightarrow{C'C}$ is the direction of the instantaneous acceleration of the planet in its orbit. Law 2, then, effectively states that the acceleration of each planet in its orbit is at all times directed towards the sun.

The beginnings of a translation of Kepler's third law into terms of accelerations were made by Newton as early as 1666. Newton was aware that the orbits of the planets about the sun did not differ very markedly from 'perfect' circles: Kepler's ellipses were, for all the then known planets, of small eccentricity. So, in order to get the 'feel of the problem', Newton made the approximation that the orbits were indeed circles, and that the motion in each was with uniform speed (as Law 2 would in consequence predict). Suppose, then, that two planets are moving in circular orbits of radii r_1 and r_2, respectively, and that their periodic times are T_1 and T_2. The centripetal accelerations of the planets being given by

$$a_1 = \left(\frac{2\pi}{T_1}\right)^2 r_1 \qquad a_2 = \left(\frac{2\pi}{T_2}\right)^2 r_2 \quad \text{(see equation 4.19),}$$

we have $\dfrac{a_1}{a_2} = \left(\dfrac{T_2}{T_1}\right)^2 \dfrac{r_1}{r_2}.$

But, by Kepler's third law, we have

$$\left(\frac{T_2}{T_1}\right)^2 = \left(\frac{r_2}{r_1}\right)^3,$$

149 Universal Gravitation: the Astronomical Evidence

therefore, we conclude that

$$\frac{a_1}{a_2} = \left(\frac{r_2}{r_1}\right)^2.$$
7.9

Considering this highly idealized case, we are led to the tentative view that as between the various planetary orbits the centripetal acceleration in no way depends on the physical characteristics of the planet occupying the orbit, but merely on the orbital radius – varying inversely as the square of that quantity throughout the solar system generally. This is a view of great simplicity, and therefore of intrinsic attractiveness (see p. 135). From the first, Newton believed that it held the truth of the matter, but for many years proof eluded him. In order to establish it, it was necessary to show that the characteristics of the single (non-circular) orbits of the individual planets issued directly from its premises. More specifically, it was necessary to show that motion in an ellipse, with speed varying in such a way that the instantaneous acceleration is at all times directed to one focus of the ellipse, necessitates an inverse-square law of acceleration as determining condition. It was proof of this proposition which eluded Newton – and many another mathematician of the first rank – until 1685.

A little before that time the whole problem of the interpretation of Kepler's laws had become one of great interest to the Fellows of the Royal Society: it had been discussed at length by Wren, Hooke and Edmund Halley (1656–1742), meeting in London. Eventually, in January 1684, Hooke claimed to have a full solution. The others waited until August for Hooke to substantiate his claim, but nothing came of it, and Halley journeyed to Cambridge to consult Newton on the matter. Newton was emphatic that, some seven or eight years previously, he had proved to his own satisfaction that if, under an inverse-square law of acceleration directed towards a fixed point, a particle moves in a closed orbit, then the orbit is an ellipse and the point is one focus of it (this is the converse of the proposition that we enunciated in the last paragraph). Unfortunately, Newton had mislaid his earlier proof, but, being challenged in this way, he reconstructed it and sent it to Halley in November. Halley visited Cambridge again in December 1684, and found that Newton had indeed progressed much farther than to reproduce his earlier proof. (To this meeting the eventual publication – if not the main incentive to Newton's writing – of *Principia* may justly be ascribed.) Sometime in the following year, and certainly before April 1686, Newton had two independent proofs of the primary proposition as well. Both were included in the published work.

It would be beyond the scope of this book to give either the proof of the primary proposition, as enunciated in the last paragraph but one, or the proof of its converse. The latter is to be found in almost all standard texts on applied mathematics, in the section on 'central orbits'; the former is to be found in some of them. Suffice for our purposes to draw attention to one aspect of the kinematical problem which Newton emphasized in one of his proofs of the primary proposition. We have already discussed one particular type of elliptical motion in the context of the superposition of linear simple harmonic motions about the same origin, in mutually perpendicular directions and of the same period (p. 60). In that case the speed in the orbit is such that the point in the auxiliary circle moves with constant speed throughout. It is easy to show that in those circumstances the acceleration of the particle in the ellipse is at all times directed towards the centre and is directly proportional to the distance from the centre (the component accelerations are proportional, respectively, to the projections of the central distance of the particle on the lines in which the superposed simple harmonic motions are effective). The law of speed variation in the elliptical orbits of the planets is entirely different from this. In one of his proofs, Newton deduced the (unknown) law of acceleration variation in the planetary orbit from the (known) law of variation in a geometrically identical orbit of the other type, using a subtle and very elegant geometrical argument. We draw attention to this fact of history to emphasize, in our turn, the clear distinction between these two types of motion in an ellipse. Such an orbit is possible either when the acceleration of a particle is directed towards a fixed point and is proportional to its distance from that point (when the point becomes the centre of the ellipse), or when the acceleration is directed towards a fixed point and is inversely proportional to the square of the distance (when the point becomes one focus of the ellipse).

Following this digression, let us now summarize the situation as it appeared to Newton in the year 1685. He had shown unambiguously that to accept Kepler's laws (at least as an approximate description of reality) was to accept just one simple conclusion: any planet moving 'naturally', 'under the influence of' the sun, at all times has an acceleration directed towards the sun, the measure of the acceleration being given by the product of the inverse square of the measure of the distance of the planet from the sun and the measure of some other parameter which is the same for all planets, at whatever distance they may be. We may express this conclusion formally as follows:

$$a = \frac{\gamma}{r^2}. \hspace{4cm} 7.10$$

In equation 7.10, a and r represent the measures of acceleration and distance, in traditional notation, and γ is the measure of the 'other parameter' to which we referred. (Incidentally, in relation to Kepler's third law, Newton had shown that, for formal precision, the original phrase 'their mean distances from the sun' should be replaced by 'the major diameters of their orbits'.)

On the basis of his newly formulated definitions of (inertial) mass and force, Newton would naturally be tempted to transform equation 7.10, representing the kinematical situation which Kepler's laws disclosed, into the dynamically significant equation

$$F = \gamma \frac{m}{r^2}. \hspace{4cm} 7.11$$

151 Universal Gravitation: the Astronomical Evidence

According to equation **7.11**, each planet experiences a force directed towards the sun, inversely proportional to the square of the distance and directly proportional to its own mass. In his own third law of motion, Newton would recognize his confident assertion that this force must be matched by an exactly equal force, experienced by the sun and directed towards the planet. At this point, apparently, his remarkable intuitive powers again took command. If the force acting from the sun on the planet is proportional to the mass of the planet, then, almost of necessity, the equal and oppositely directed force acting from the planet on the sun must be proportional to the mass of the sun. In this moment of insight, effectively, equation **7.11** had taken on the form

$$F = G \frac{mM}{r^2},$$
7.12

M representing the measure of the (inertial) mass of the sun, and G the measure of a new parameter, the 'gravitational constant', as we define it today.

There are several comments that must be made in elaboration of the law of universal gravitation, as it is exemplified by equation **7.12** 'deduced' on the basis of Kepler's laws. In the introduction to this chapter we stated the law in its generality (p. 141). Then we raised the question (p. 144) whether it is not more significant to state it otherwise, introducing the concept of gravitational mass, as a physical quantity distinct from inertial mass though intimately related to it. Following that suggestion we were led to a definition of gravitational mass symbolically in terms of equation **7.1**. (We might have inserted another 'multiplying constant' in equation **7.1** – the same as, or different from, that represented by G in equation **7.12** – and so have obtained an alternative definition of this quantity.) We do not claim that our subsequent discussion of the problem resolved the dilemma of 'the two masses': the question of the reality of the distinction remains, for us, an open one. We merely insist that, in its mode of derivation from observation, equation **7.12** refers unambiguously to the inertial masses of the bodies concerned.

Our next comment is of a different character. When we quoted the law of universal gravitation at the beginning of this chapter, we left no doubt regarding its claim to universality: 'Between each pair of particles in the universe ...', we wrote. If we do not allow ourselves to be dazzled by the brilliance of Newton's 'deduction' of equation **7.12** from Kepler's laws, we shall almost certainly recognize an aspect of logical inconsistency in its subsequent generalization. We shall come to realize that the motions of the planets have been interpreted on the assumption that each planet is acted on by one force, exclusively – the force of gravitational attraction between that planet and the sun – and, when apparent order has been imposed on the facts of experience on the basis of that assumption specifically, we have jettisoned it, without comment or qualm, and blandly asserted (on the evidence of our previous success!) that every particle in the universe attracts every other particle (every planet, therefore, every other planet) with the same kind of force. Possibly, we have already lost sight of the fact that the expression of Kepler's laws – and that of Newton's interpretation of them – is irrevocably conditioned by the 'particle approximation' to which we referred at the outset (p. 148). Pragmatically (and in sober numerical estimation as well), the particle approximation has proved to be justified, but the real situation could have been other than it is – the validity of the particle approximation is entirely uncovenanted. In the same way, it is entirely uncovenanted that the mass of the sun is so very much greater than the mass of any of the planets (more than a thousand times greater than the mass of the largest of them, as we now know) that the forces of gravitational attraction as between one planet and another are negligible compared with the force with which either is attracted to the sun. Herein, again, lies the basis of the pragmatic justification of the seemingly illogical transition from equation **7.12** to the universal law. The solar system in which we live is obviously, through the accident of its configuration, one of the simpler types of system from which to win understanding – at least to a Newton.

In closing the last paragraph as we have done, we may have given the impression that with the interpretation of Kepler's laws, and the (logically insecure) formulation of the law of universal gravitation, based on their evidence alone, Newton had written the last word on the subject of gravitation in the context of planetary motion. The facts are entirely otherwise. Kepler's laws are merely statements of empirical regularity valid as good approximations to the truth, and nothing more – as we have already admitted (p. 151). The crowning success of Newton's law of gravitation was not that it provided a satisfactory correlation of these over-all regularities, but that it has provided in detail, and over a long period, equally satisfactory 'explanations' of the major discrepancies that have arisen as between later observations and the predictions of Kepler's laws. Newton laid the foundation of this process of continual reassessment in Book 3 of *Principia*. It may be regarded as having culminated in the theoretical work of Pierre Simon, Marquis de Laplace (1749–1827), whose *Mécanique céleste*, published in five volumes between 1799 and 1825, earned for him the title of 'the Newton of France'.

7.3   Universal gravitation: the system 'earth–apple–moon'

Tradition has it that Newton's attention was directed towards the possibility of a theory of universal gravitation by the fall of an apple, as he was walking in the orchard of the manor house at Woolsthorpe, in Lincolnshire, in the autumn of the year 1666. Throughout that summer, as in the previous year, students had been dismissed from their Cambridge colleges because of the plague, and Newton had, indeed, spent the time largely in his Lincolnshire birthplace. Moreover, when he came to survey his achievements, as an old man, he had written, 'In the same year [1666] I began to think of gravity extending to the orb of the moon ... [I] compared the force requisite to keep the moon in her orb with the force of gravity at the surface of the earth, and found them to answer pretty nearly ... in those days I was in the prime of my age for invention, and minded mathematics and philosophy more than at any time since.' Tradition, then, enshrines a symbolical, if not a literal, truth. The title chosen for this section (and repeated from an earlier allusion – p. 148) suggests as much. We shall interpret it as covering our discussion of the relevance of the law of universal gravitation to sub-lunar phenomena generally. Brief reference has already been made to one aspect of this problem in the last section.

In 1609 Kepler published the first and second of his laws of planetary motion (p. 40). In the same year, Galileo, having heard in Venice of the invention of Lippershey in Holland, returned to Padua and, working out the principles for himself, constructed his first telescope. In 1619 Kepler's third law was published. In that year Jeremiah Horrocks was born, near Liverpool in England. In each case a significant component of knowledge was thereby added to the information upon which Newton was able to reflect when he began to think of terrestrial gravity as 'extending to the orb of the moon'. Before he died at the age of twenty-one, Horrocks had shown how Kepler's first law could be amended to accommodate the moon's motion as it is seen from the earth. In that moving frame of reference the orbit of the moon is in an ellipse of periodically varying eccentricity and slowly rotating major axis. (In Book 3 of *Principia* Newton was to show that this description validly summarized the essential predictions of his gravitational theory.) For his part, in January 1610, Galileo discovered the four major satellites of Jupiter: the earth was not alone in the solar system as possessing a moon. (Newton was later to demonstrate the validity of Kepler's third law, in particular, in respect of the Jovian satellites.)

In 1666, as we have already seen (p. 149), Newton was surveying the evidence concerning gravitational phenomena using the simplest approximations. For the problem on which we are now concerned one of these approximations (or assumptions) was more difficult to justify convincingly than the other. Even taking account of the work of Horrocks, it was not significantly more 'dangerous' to regard the moon's orbit around the earth as circular (and its speed in the orbit as constant), as Newton did, than to regard the orbits of the planets around the sun as circular for the purposes of the assessment of Kepler's third law as we have already described (p. 149). But in relation to the apple – that is, to the value of the acceleration due to gravity at the surface of the earth, it was another question altogether. In the calculation which '[answered] pretty nearly', Newton assumed that the whole earth attracts a body at its surface 'as if the whole mass of the earth were concentrated at its centre'. For that assumption he had no justification at the time. (It may be remarked that he did not publish his calculation, anyway.) Let us follow the calculation, then, without scruple, and return, as Newton did, to the justification of the underlying assumption when we see where it has led.

Let us suppose that the moon's orbit is circular and of radius R. Let T be the periodic time of this orbital motion. Then the earth-directed acceleration of the moon is given by $(2\pi/T)^2 R$. Let us denote the radius of the earth (assumed spherical) by r, and the acceleration due to gravity at the surface by g. Then, if the inverse-square-law distance variation applies to these two accelerations, and if we accept the assumption which is the only serious cause of doubt, we shall expect to have

$$\left(\frac{2\pi}{T}\right)^2 \frac{R}{g} = \left(\frac{r}{R}\right)^2,$$

or $\qquad T = \frac{2\pi R}{r} \left(\frac{R}{g}\right)^{\frac{1}{2}}.$ $\qquad\qquad$ 7.13

When Newton stated that his calculation answered pretty nearly, he meant that the value of T as calculated from equation 7.13 in terms of the accepted values of g, and of the two lengths entering into the equation, agreed with the observational value within some 16 per cent. Not everyone, perhaps, would have regarded this answer as 'pretty near', but such was the judgement of genius – and the outcome was equally surprising. Three years later, in 1669, a new determination of the radius of the earth was made by Jean Picard. When Picard's value for r was inserted in the equation for T, the discrepancy was reduced to a mere 1·6 per cent. Still Newton did not publish. However, in 1685, engaged in writing *Principia*, he returned to the problem. Then, he succeeded in justifying his doubtful assumption – and the whole debate on terrestrial and solar gravitation fell into order, and the claim of validity for the law of universal gravitation was enormously strengthened.

As a last example of Newton's consummate skill in devising geometrical proofs of theorems in physics, we give in what follows, essentially as it occurs in proposition 71 of Book 1 of *Principia*, the proof that originally was lacking. We shall show, as Newton did, that a uniform spherical body attracts an exterior particle as if the whole of the mass of the body is concentrated at its centre. In relation to the problem of the apple, we merely remark, in parenthesis, that Newton was not committed to the assumption that the earth is, in fact, a uniform spherical body (which we know it is not); as the form of the proof will make clear, all that it is necessary to assume is that the density is a function of the distance from the centre, exclusively. In relation to the apple, the earth may indeed be onion-like.

Let us suppose, then, that in Figure 25 the outer circle represents a central section of a uniform spherical body A, of material of density ρ, and of radius R_0, C being the centre of the sphere. Let a particle of mass m be situated at P, a point external to A. Let us consider, first, the resultant gravitational force acting on this particle from the material included in the spherical shell A′, concentric with A, having internal and external radii R and $R + \Delta R$. From considerations of symmetry, it must be clear that this resultant force acts in the direction $\overrightarrow{PC}$. The force due to an element of the shell about Q, however, acts on the particle along $\overrightarrow{PQ}$. Suppose, for precision, that Q is on the inner surface of A′; let $PQ = r$, $PC = d$, and let P′ be the inverse point of P with respect to

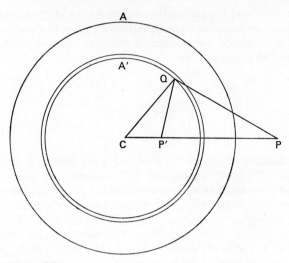

Figure 25

A' (the definition of the inverse point requires that $CP'.CP = R^2$). $P'Q$ having been joined, let us consider a cone of small apical solid angle $\Delta\omega$ having its apex at P' and with P'Q as axis. Let this cone intersect the inner surface of A' in an area $\Delta\sigma$. Then the volume of A' included within the cone is $\Delta\sigma\,\Delta R$ (neglecting small quantities of higher order) and the force of mutual attraction of the particle at P and the material of the sphere which is included within this element of volume is of magnitude

$$G\frac{m\rho\,\Delta\sigma\,\Delta R}{r^2},$$

G being the measure of the gravitational constant, as in equation **7.12**. As we have stated, this force acts on the particle at P in the direction $\overrightarrow{PQ}$: its component in the direction $\overrightarrow{PC}$ is given by

$$\Delta^2 F = G\frac{m\rho\,\Delta\sigma\,\Delta R}{r^2}\cos\theta,\qquad\qquad\text{7.14}$$

where, in this context, the symbol Δ^2 refers to a small quantity of the second order, and $\angle QPC = \theta$.

Now, according to the definition of the inverse point,

$$\frac{CP'}{CQ} = \frac{CQ}{CP},$$

moreover, these two ratios are the ratios of the lengths of pairs of lines which include the same angle, $\angle QCP$. We conclude, therefore, that $\triangle CQP'$ and $\triangle CPQ$ are similar triangles, and, in particular, that

$$\angle P'QC = \theta,$$
$$\frac{P'Q}{CQ} = \frac{QP}{CP}.$$

If we write $P'Q = R'$, the second result may be rewritten

$$\frac{R'}{R} = \frac{r}{d},$$

then equation **7.14** becomes

$$\Delta^2 F = G\,\frac{m\rho R^2\,\Delta R}{d^2}\,\frac{\Delta\sigma\cos\theta}{R'^2}.$$

Also, because of the first result, we have

$$\Delta\omega = \frac{\Delta\sigma\cos\theta}{R'^2},$$

thus, finally, we obtain

$$\Delta^2 F = G\,\frac{m\rho R^2\,\Delta R}{d^2}\,\Delta\omega. \qquad\qquad \textbf{7.15}$$

Equation **7.15** gives the component of force in the direction $\overrightarrow{PC}$ acting on the particle at P from the material in the element of volume of A' about Q which we have already identified. As the form of the equation shows, this force component is directly proportional to $\Delta\omega$. This being the case, we may sum directly over the whole solid angle about P' to obtain the component of force on the particle at P due to the complete spherical shell A'. The result is $(\sum \Delta\omega = 4\pi)$

$$\Delta F = G\,\frac{m\rho 4\pi R^2\,\Delta R}{d^2};$$

or, alternatively, $\quad \Delta F = G\,\dfrac{m\rho\,\Delta V}{d^2}, \qquad\qquad \textbf{7.16}$

where ΔV is the volume of A'. On the basis of equation **7.16**, we may now sum over the volume of the whole sphere A, if ρ is constant throughout as we have already assumed. We obtain in this way

$$F = G\,\frac{m\rho V}{d^2}$$

V being the volume of A. Alternatively, again,

$$F = G \frac{mM}{d^2},$$ 7.17

if M is the total mass of the spherical body concerned. Obviously, equation 7.17 gives the actual force acting on the particle at P in the direction $\overrightarrow{PC}$. It is precisely as if the whole mass of the body A had been concentrated at its centre C.

Further reading

A. B. Arons, *Development of the Concepts of Physics* (Chapter 15), Addison-Wesley, 1965.

H. Butterfield, *The Origins of Modern Science, 1300–1800* (Chapter 8), Bell, 1949.

S. Toulmin and J. Goodfield, *The Fabric of the Heavens*, Harper & Row, 1961.

Chapter 8
Relativity

8.1 Newtonian relativity (I)

In our discussions of Newton's laws of motion in Chapter 6, and of the law of universal gravitation in Chapter 7, we have sometimes been less than precise regarding the kinematical frames of reference in relation to which our statements have been made. To some degree such vagueness of reference was excusable, if only because we had so much else to discuss. In this section, however, we seek to make amends. Precision in this matter, as we shall see, is fundamentally important.

In general, the results of our previous discussions may be summarized in the statement that except in extreme situations (which, it must be admitted, we have not yet identified specifically) the results of laboratory experiments and astronomical observations, alike, are found to be consistent with Newton's laws. At first sight, this statement appears unexceptionable, and nothing that we have said so far casts serious doubt on it. Indeed, we still intend to adhere to it, but we have to make one important proviso. The human observer need not always take the immediate evidence of his senses at its face value: he must be sufficiently aware of the peculiarities of his local situation to be able to rationalize his sense data appropriately. Let us take a particular example, in illustration.

In 1851 Jean Bernard Léon Foucault (1819–68) suspended a polished brass sphere weighing 5 kg by a steel wire about 2 m long from an iron clamp in the roof of his cellar. He set this simple pendulum in uniplanar motion with great care, by first securing the bob in a displaced position by means of a thin thread, and then, when all movement had ceased, burning the thread. Such a heavy pendulum, set in motion in this way, continues to oscillate for a considerable time. Foucault observed that the plane of oscillation of his pendulum slowly changed direction, rotating in a clockwise sense as seen by an observer looking downwards. Shortly afterwards, he transferred his experiments to the Panthéon, in Paris, using a bob of 28-kg weight and a suspension of 67 m. In this way he was able to observe the free oscillations of the pendulum for periods of some six hours at a time, and to show that the rate of rotation of the plane of oscillation was independent of the initial azimuth of that plane and such that the period of complete rotation was about thirty-two hours.

According to the 'immediate evidence of the senses' the motion of Foucault's pendulum bob had two obvious components: an oscillatory motion in a plane, and a transverse motion involving apparent revolution about the vertical axis. Each of these motions involves a varying acceleration. The acceleration in the oscillatory motion can be understood in detail in terms of Newton's third law. The force which corresponds to it can itself be analysed into two components, one the resultant of all the gravitational forces acting between the constituent particles in the bob and the constituent particles of the earth (according to the law of universal gravitation) – and the other the resultant of the 'elastic' forces arising (see p. 117) in the slightly deformed region of the bob where it is attached to the (slightly deformed) wire which constitutes the suspension. The other (admittedly very much smaller) acceleration cannot be similarly understood. The force which corresponds to this acceleration cannot be identified as one member of 'a pair of equal and oppositely directed forces acting between two material particles': its 'presence' appears to contravene the fundamental law of nature on which Newton's whole dynamical scheme is based.

Here, however, the 'human observer', if he 'is sufficiently aware of the peculiarities of his local situation', may well realize that the earth on which he is situated is rotating about its polar axis with reference to the 'fixed' stars with a period of some twenty-four hours. The component of this motion of steady rotation, taken about an axis through the centre of the earth and his point of observation (in Paris) – that is the component of rotation about the vertical axis through the point of suspension of Foucault's laboratory pendulum – is a uniform rotation with a period of about thirty-two hours (and an anticlockwise sense as seen from above). If he recognizes the significance of this fact he will see that in relation to directions fixed with respect to the stars, the plane of oscillation of his pendulum has no resultant rotation. With respect to this new frame of reference there is no need to postulate the existence of any unpaired forces: Newton's laws are in no way contravened. Indeed, it is as if the distant stars were the guarantors of the validity of Newton's laws in respect of an experiment carried out in a cellar in Paris!

The successful realization of the conditions necessary for the demonstration of the Foucault pendulum effect requires ingenuity and care, but the situation in which this effect occurs is in no way an 'extreme' situation. In quite ordinary circumstances, then, Newton's laws are not 'literally true' in respect of macroscopic phenomena observed in a laboratory on the surface of the earth. On the other hand, they are true, as far as experiment can decide, in respect of the same phenomena, if the rotation of the earth with respect to fixed stars is suitably 'allowed for'.

In relation to the motion of the planets, if we take the heliocentric view of Copernicus as simplified by Kepler, the laws are validated to a high degree of approximation when these motions are referred to coordinate axes fixed in the centre of mass of the sun and orientated with respect to the stars. They are even more closely validated when the origin is transferred to the mass centre of the solar system as a whole (this point is never more than a fraction of the sun's radius outside its surface). When discrepancies appeared in relation to the motion of Uranus in 1820, the upshot was the prediction of the existence of a 'new' planet in 1845, and its discovery in the following year. The motion of this planet (Neptune), in its turn, showed slight 'irregularities', and eventually Pluto was discovered in 1930. Only one slight irregularity failed to yield to 'Newtonian' treatment. In 1845 U.J.J.Leverrier (1811–77) reported that the perihelion of the orbit of Mercury (the planet nearest to the sun) appeared to be advancing more rapidly (by some thirty-five seconds of arc per century) than could be directly accounted for. For seventy years this irregularity (which subsequent observation confirmed, and 'corrected' to forty-three seconds per century) defied all rational explanation. Eventually, it required an essentially new theory of gravitation for its understanding. That was Einstein's general theory of relativity (1915) – but we shall not be able to extend our discussion as far as to develop that theory in this book (see p. 189).

Recoiling, then, from a premature discussion of twentieth-century relativity, let us summarize the Newtonian position. It will be convenient to start from Newton's own statement – for it is clear that he had given considerable thought to the problem. In his fifth corollary, which we have not quoted hitherto, he wrote, 'Bodies inclosed in a given space have the same motions relatively to one another, whether that space be at rest, or be moving uniformly in a straight line without rotation.' Regarded literally, this statement is not very helpful: the notion of a frame of reference ('space') absolutely at rest is one that we have long since abandoned as meaningless (p. 18). Moreover, Newton himself confessed to inability to identify such a space: 'It is possible that in the remote regions of the fixed stars, or perhaps far beyond them, there may be some body absolutely at rest, but impossible to know ... absolute rest cannot be determined from the position of bodies in our regions.' However, Newton's fifth corollary is not fundamentally meaningless, if properly interpreted. We shall interpret it as follows. Let there be two spaces, one in uniform rectilinear motion with respect to the other. Then the accelerations of bodies in relation to the second space are identical with their accelerations in relation to the first. Consequently, if Newton's laws are valid for observations made by an observer at rest in the first space, they are likewise valid for observations similarly made in the second space. According to the primary notions of space and time which Newton instinctively accepted, this proposition is indeed self-evident. Let us see why that is so, and what hidden assumptions are involved in this point of view.

Let OX, OY, OZ (Figure 26) be rectangular axes such that observations made with respect to these axes (and an agreed system of time measurement) are consonant with Newton's laws of motion. Let O' be a point in OX, and let $O'X'$, $O'Y'$, $O'Z'$ be rectangular axes through O', such that $O'Y'$ is parallel to OY, and $O'Z'$ is parallel to OZ. Then $O'X'$, of necessity, is collinear with OX. Let the figure represent the position at an arbitrary time t, and let us suppose that, for all values of t, $OO' = vt$, where v is a constant quantity. Obviously, v is the measure of the velocity of rectilinear motion of the second set of axes with respect to the first. (Hereafter we shall denote these two frames

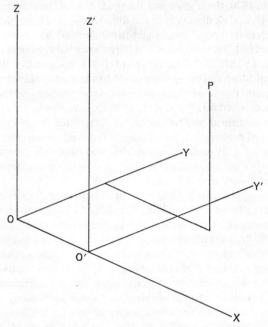

Figure 26

of reference ('spaces') by S and S', respectively.) Suppose that there is a particle situated at P, and that its coordinates with respect to S and S' are (x, y, z) and (x', y', z'). Clearly,

$$\left.\begin{aligned} x' &= x - vt, \\ y' &= y, \\ z' &= z. \end{aligned}\right\} \tag{8.1}$$

Let us make the assumption that the same system of time measurement is available to observers at rest in S and S'; then we have, differentiating,

$$\left.\begin{aligned} \dot{x}' &= \dot{x} - v \\ \dot{y}' &= \dot{y} \\ \dot{z}' &= \dot{z} \end{aligned}\right\} \quad \text{and} \quad \left\{\begin{aligned} \ddot{x}' &= \ddot{x}, \\ \ddot{y}' &= \ddot{y}, \\ \ddot{z}' &= \ddot{z}. \end{aligned}\right.$$

We conclude that the acceleration (though not the velocity) of P is the same for both observers – and because Newton's laws deal with accelerations, exclusively, that these laws are valid for observations in S', if they are valid for observations in S, as we have originally postulated.

In the argument that we have just concluded we have shown how the statement of Newton's Corollary 5 follows by direct deduction from his basic assumptions regarding the character of space and time. As we have presented the argument, we have drawn attention to one assumption, in particular, which might otherwise have been overlooked – that a common system of time measurement is, in principle, always available to two observers moving with different velocities (in respect of any arbitrarily chosen reference frame). This assumption, trivial enough in Newtonian philosophy, has proved, in the up-shot, to be the weakest link in the chain of argument that we have just presented. We shall see, in the next section, how experience of certain phenomena (in 'extreme situations') has led to its modification; for the present let us return to consider the implication of the last two words, 'without rotation', which Newton inserted – not without deliberate intention – in Corollary 5.

Suppose that two rectangular reference frames, S and S', are completely indistinguishable at $t = 0$. Let S' be rotating with constant angular speed ω about $OZ(OZ')$. Then, taking an arbitrary point P as before, the transformation equations expressing the position of P in S' at time t, in terms of its simultaneously observed position in S, are

$$x' = x \cos \omega t + y \sin \omega t,$$
$$y' = y \cos \omega t - x \sin \omega t,$$
$$z' = z.$$

From these equations, on the basis of our previous assumption regarding the availability of a common system of time measurement, we have, on differentiation,

$$\dot{x}' = \dot{x} \cos \omega t + \dot{y} \sin \omega t + \omega y',$$
$$\dot{y}' = \dot{y} \cos \omega t - \dot{x} \sin \omega t - \omega x',$$
$$\dot{z}' = \dot{z}.$$

and, after some reduction,

$$\left. \begin{aligned} \ddot{x}' &= \ddot{x} \cos \omega t + \ddot{y} \sin \omega t + 2\omega \dot{y}' + \omega^2 x', \\ \ddot{y}' &= \ddot{y} \cos \omega t - \ddot{x} \sin \omega t - 2\omega \dot{x}' + \omega^2 y', \\ \ddot{z}' &= \ddot{z}. \end{aligned} \right\} \qquad \textbf{8.2}$$

Each of the first two of equations **8.2** contains four terms in its right-hand member. If only the first two terms in each case were significant, we should have $\ddot{x}'^2 + \ddot{y}'^2 = \ddot{x}^2 + \ddot{y}^2$ – and the magnitude of the acceleration of P in S' would be the same as the magnitude of its acceleration in S. Moreover, if two particles appeared in S to have accelerations which were oppositely directed, the same two particles would appear to have oppositely directed accelerations in S'. On the basis of the first two terms in the right-hand members of the first two of equations **8.2**, therefore, the validity of Newton's laws in respect of both S and S' would be a logical possibility; taking equations **8.2** as they have in fact emerged from our analysis, however, that possibility is specifically discounted. Newton's laws of motion cannot be valid, on the basis of Newton's own assumptions, in respect of observations made in each of two frames of reference one of which is rotating with respect to the other. Hence the additional emphasis of the two words, 'without rotation', which are strictly redundant in Corollary 5.

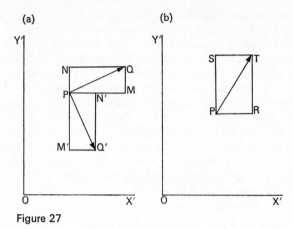

Figure 27

Newton might also have added 'or linear acceleration' to provide final emphasis to his statement of the corollary. The words would have been equally redundant – and, indeed, he did not add them. However, the reader might still be encouraged to examine this possibility for himself. If, instead of equations **8.1**, he starts from the equations

$$x' = x - \tfrac{1}{2}at^2, \quad y' = y, \quad z' = z,$$

he will be dealing with frames of reference one of which is accelerated with respect to the other with linear acceleration a. He will quickly discover that Newton's laws cannot be valid for observations made in each of these frames.

Let us return for a moment to equations **8.2**, before we summarize the situation regarding Newtonian relativity. Let us suppose that S is a frame of reference in respect of observations in which Newton's laws are valid. Then accelerations as observed in S' do not lead to the identification of forces which are associated in equal and oppositely directed pairs effective as between pairs of material particles. From our position of 'superior insight' (having, on the basis of our equations, a point of observation in each reference frame), we recognize that the components of acceleration represented by the third and fourth terms in each of the first two of equations **8.2** represent the supernumerary, or 'fictitious', accelerations. The third terms in each of these equations describe components of acceleration which depend upon the velocities of particles as observed in S', the fourth terms describe components of acceleration which depend on their instantaneous positions. These various components are represented 'to scale' in Figure 27(a) and (b).

In Figure 27(a), $\overrightarrow{PQ}$ is a vector representing the resultant of $\dot{x}'$ and $\dot{y}'$, the components, parallel to OX' and OY', respectively, of the instantaneous velocity of P in S'. $\overrightarrow{PM}$ and $\overrightarrow{PN}$ represent these components individually.

According to the third terms in the equations, a component of acceleration proportional to PN is effective in the direction $\overrightarrow{OX'}$ and a component proportional to PM is effective in the direction $\overrightarrow{Y'O}$. These components of acceleration are represented by $\overrightarrow{PN'}$ and $\overrightarrow{PM'}$. Obviously, from the geometry of the figure, their resultant, represented by $\overrightarrow{PQ'}$, has a direction at right-angles to $\overrightarrow{PQ}$. If v' is the magnitude of the velocity represented by $\overrightarrow{PQ}$, the magnitude of the acceleration represented by $\overrightarrow{PQ'}$ is $2\omega v'$ (see equations 8.2). It will be noted that the direction of this acceleration is at right angles to both the direction of v' and the axis of ω (and is in sense opposed to that of the rotation of S' about OZ in S). This component of supernumerary acceleration in a rotating frame of reference was first treated systematically by G. G. Coriolis in 1844, and generally goes by his name. Essentially, the motion of Foucault's pendulum (p. 159), as observed on the rotating earth, exhibits the effect of this acceleration.

In Figure 27(b), the components of the supernumerary acceleration described by the fourth terms in equations 8.2 are represented by $\overrightarrow{PR}$ and $\overrightarrow{PS}$. According to the equations, these components are proportional, respectively, to x' and y', the Cartesian coordinates of P in the plane $X'OY'$. Obviously, then, the resultant, represented by $\overrightarrow{PT}$, is directed along $\overrightarrow{OP}$ produced, and is of magnitude $\omega^2 r$, if $OP = r$. This supernumerary acceleration is directed away from the axis of rotation of S': it is usually referred to as the centrifugal component. A body permanently at rest on the surface of the rotating earth must be acted on by Newtonian forces appropriate in magnitude and direction to give it acceleration of amount $\omega^2 r$ towards the polar axis of the earth. Regarded in the purely local situation, such a body has no acceleration. According to equations 8.2 this observation is accounted for ($\ddot{x}' = \ddot{y}' = \ddot{z}' = 0$) in that the first two terms in the first two of these equations represent this centripetal acceleration (due to the 'real' forces), the fourth terms in these equations represent the oppositely directed centrifugal acceleration of the same magnitude. The third terms, of course, are zero ($\dot{y}' = \dot{x}' = 0$).

We are now in a position to summarize. It is the essential principle of Newtonian relativity that there is no distinction, dynamically, between two frames of reference in uniform rectilinear motion one with respect to the other. If by any means one frame of reference may be identified (no recipe is given for this identification) in relation to which observations are precisely consonant with the 'requirements' of the Newtonian laws, then an infinity of other such frames may be specified (frames having uniform rectilinear motion – of whatever velocity – with respect to the first) which have the same property. In the hopeful belief that this infinitely numerous set of reference frames is no mere figment of an over-sanguine imagination, a special name is given to its members. They are referred to as *inertial frames*. Other frames of reference, according to this view, are not inertial frames. These are the infinity of frames which are in any way accelerated in relation to frames of the first set. In Newtonian relativity the equations of transformation, as between one inertial frame and another, are our equations **8.1** – supplemented by the all-important equation for time transformation, $t' = t$. If the premisses on which this view of the kinematical structure of the real world is based are open to criticism, perhaps the most vulnerable premiss is that which asserts that a common system of time measurement is, in principle, always available to two observers in relative motion, one with respect to the other (p. 163).

8.2 Newtonian relativity (II): the propagation of light

Instinctively, in common experience, we act on the assumption that the evidence which our eyes present to us, even of relatively distant events within our local scale of relevance, is 'immediate' evidence in a way in which the evidence of our ears is not. Through long familiarity with the phenomenon, we have come to associate the flash of lightning and the subsequently heard peal of thunder as components of a single physical process which have become separated in observation only because the atmospheric disturbance which is sound 'takes time to travel' from its place of origin to where we are. In common experience, we find no need to make a similar interpretative adjustment to those of our sense data which come to us directly through the sense of sight. In our everyday lives, we encounter no inconsistencies if we assume that light travels instantaneously from source to observer. Evidence that this assumption is untenable, on the astronomical scale, was first provided by Ole Roemer (1644–1710) only twelve years before Newton's *Principia* was published. In 1675 Roemer interpreted a small periodic variation of the times of eclipse of the first Galilean satellite of Jupiter, as observed from the earth, as arising from the finite velocity of propagation of the light by which these events are observed. The distance from the earth to Jupiter certainly varies periodically (with a period of 399 days); any delay in the terrestrial observation of Jovian events, therefore, must vary with this period. By the time that Newton came to write the additional matter for the second edition of his *Opticks* (1717), Roemer's interpretation of the astronomical evidence had been generally accepted; new and more accurate observations had been made, and it was possible for Newton to state, as if it were an established fact, 'Light moves from the Sun to us in about seven or eight Minutes of Time.'

Stating 'the essential principle of Newtonian relativity', in our summary at the end of the last section, we were careful to use the qualifying adverb 'dynamically': 'there is no distinction, dynamically, between two frames of reference in uniform rectilinear motion one with respect to the other'. The question immediately arises, whether, according to this formulation, the scope of the principle is wide enough to encompass the phenomenon of light propagation. Newton would certainly have insisted that it is. 'And are not the Rays of Light very small Bodies ... which by their attractive Powers, or some other

Force, stir up Vibrations in what they act upon ...' he wrote. His light corpuscles were clearly conceived as interacting with the constituent particles of gross bodies with (Newtonian) forces: they were subject to the laws of motion as all other particles were. Their motion through interstellar space was envisaged as possibly determined by their interactions with the (even smaller) particles of an all-pervading 'Medium': 'And so if any one should suppose that *AEther* (like our Air) may contain Particles which endeavour to recede from one another (for I do not know what this *AEther* is) and that its Particles are exceedingly smaller than those of Air, or even than those of Light. . . .'

Newton had no objective evidence for his 'particles of light' – and assuredly none for his 'particles of aether' – but it was natural that he should make models of reality, to provide a ground on which his imagination could work, and this was the type of model that he favoured. On the basis of this model, the principle of Newtonian relativity should be equally applicable to observations made on the motions of particles of light as to observations made on the motions of material bodies. The inclusion of the phenomena of light propagation within the scope of the principle would not change its character in any way, in particular it would not provide any infallible way of approach to the identification of a truly inertial frame. The frame of reference having its origin 'fixed' at the mass centre of the solar system, and devoid of rotation in relation to the firmament of the stars, would continue to be the nearest approach to such a frame that could be identified for practical (astronomical) use – and, in relation to any other frame in uniform rectilinear motion with respect to that frame, the velocity of propagation of light would be expected to be different from the velocity observed in the 'solar-system frame'. On the other hand, the accelerations of the light particles (effective in reflection and refraction) would be the same in the two frames – and the forces of interaction deduced from these 'observed' accelerations would be the same, also.

8.3   Relativity and the propagation of light in the nineteenth century

The falling out of favour of Newton's corpuscular theory of light can be dated almost precisely at the beginning of the nineteenth century. In 1801 Thomas Young (see p. 133), in a lecture before the Royal Society of London, developed the contrary view – which had been held in rudimentary form by both Hooke and Huygens in Newton's day – that light is essentially a wave process propagated through a continuous medium. His basic hypotheses were simply expressed. 'A luminiferous ether pervades the universe, rare and elastic in a high degree. Undulations are excited in this ether whenever a body becomes luminous.' Twenty years later, as a result of Young's own experiments on interference and diffraction, of the experiments of E. L. Malus (1775–1812) on polarization, and finally of the experimental and theoretical work of D. F. J. Arago (1786–1853) and A. J. Fresnel (1788–1827), covering this whole field of optical phenomena with great thoroughness and insight, even the most hard-ened Newtonian was converted to this new point of view. The wave theory provided a 'natural explanation' of all the effects under discussion; the cor-puscular theory had to be buttressed by adventitious assumptions even to begin to 'explain' any of them.

It was no new thing, in 1801, to postulate the existence of an all-pervading ether: as we have seen, the idea was employed, in a marginal way, by Newton himself. It certainly goes back as far as Descartes, and the eighteenth century, perhaps, was its heyday. Referring to that century, Maxwell later wrote, with obvious disapprobation, 'Aethers were invented ... to constitute electric atmospheres and magnetic effluvia ... till all space had been filled three or four times over with aethers.' However, the ether of Young and Fresnel had a more compulsive fascination for the theorist – and this fascination lasted almost to the end of the century. In 1889, in the preface to *Modern Views of Electricity*, Oliver Lodge (1851–1940) allowed himself to be overwhelmed by it: 'The existence of an ether can legitimately be denied in the same terms as the existence of matter can be denied, but only so,' he wrote.

Acceptance of the luminiferous ether was compulsive for the nineteenth-century physicist, but it was not without its difficulties. If light was a wave process, then it was essentially a transverse wave: the ether, it seemed, was incapable of transmitting a longitudinal wave of light. If elasticity was in-volved, then no known form of matter – gas, liquid or solid – had the appro-priate elastic properties. A difficulty, certainly, but from the point of view of Newtonian relativity, the more complete was the failure to comprehend the properties of the hypothetical ether in ordinary mechanical terms, the more firmly could the conviction be held that absolute space, which kinematical observations on material bodies had failed to disclose, was exhibited in the ether itself. Light, at least, so it could plausibly be asserted, was propagated in an all-pervasive 'immaterial' medium: such a medium could not con-ceivably be endowed with any sort of global motion – it must be a stagnant substrate in absolute space.

The view that we have just indicated found ample support in the develop-ment of Maxwell's electromagnetic theory, in the period between 1862 and 1872. Despite the attention devoted to it by the ablest mathematical physicists of the first half of the century, the 'elastic solid' theory of light propagation had never carried conviction. In the end, Maxwell broke completely with all attempts to invest the luminiferous ether with quasi-material properties. Instead, he postulated electrical properties for his medium. In 1856, W. E. Weber (1804–91) and R .Kohlrausch (1809–58) had made the first experi-mental determination of the ratio of the magnitudes of the electromagnetic and electrostatic units of charge. Five years previously, William Thomson (Lord Kelvin) (1824–1907) had introduced the concept of magnetic perme-ability, and the subtleties underlying the definitions of the units in the two systems were beginning to be appreciated. Independently of any theory of wave propagation, it was becoming clear that the measure of the ratio of the units of charge was the measure of a characteristic velocity of some sort. According to Weber and Kohlrausch, the magnitude of this velocity was $3 \cdot 1 \times 10^8$ m s^{-1}, and it had been noted by G. R. Kirchhoff (1824–87) how close to the free-space velocity of light this experimental figure was. To Maxwell, by 1862, the numerical coincidence was already beginning to make sense within a framework of theoretical ideas, and he was able to write, 'We can scarcely avoid the inference that light consists in the transverse undulations of the same medium which is the cause of electric and magnetic phenomena.' On this view, the propagation of light was to be described as the propagation of electromagnetic strain-energy through an immaterial ether. By 1872 the theory was essentially complete: within ten years after the death of Maxwell (1879) it had been convincingly validated by the experiments of H. R. Hertz (1857–94). Hertz was able to show that a 'dark' radiation having all the properties of visible light – reflection, refraction, polarization and (as to order of magnitude, at least) velocity of propagation – could be excited in the laboratory using large-scale electrical equipment (an induction coil and a suitable arrangement of 'radiators'). Against this background of spectacular success, even Lodge's uncritical enthusiasm could almost be excused.

Two astronomical effects, one known for a century and a half and the other only then recently discovered, gave support to the 'stagnant ether' view which Maxwell's theory appeared to demand. In 1725 James Bradley (1693–1762) had discovered that, in relation to axes fixed in and at right angles to the plane of the ecliptic (the plane of the earth's orbit around the sun), the apparent directions of the fixed stars varied periodically with a common period of one terrestrial year. In the region of the pole of the ecliptic, the stars appeared to describe very small, almost circular, orbits in the heavens, in phase one with another, and of the same angular radius (20·5 seconds of arc). In the region of the plane of the ecliptic, the apparent motion was linear of the same amplitude. This is a first-order effect (proportional, that is, to v/c, where c is the measure of the velocity of light in the neighbourhood of the observer, and v is the measure of the extreme difference of velocity of the earth in its orbit at two points diametrically opposed), and it is obviously compatible with the stagnant ether view. The second effect commonly goes by the name of one of its discoverers, Christian Doppler (1803–53). This, again, is a first-order effect: the variation of the frequency of an observed wave process with the 'line-of-sight' velocities of source and observer with respect to the medium. The effect being of the first order of small quantities, obviously only the relative line-of-sight velocity is involved when the velocities of both source and observer are small in comparison with the velocity of wave propagation. Doppler (1842), and in greater detail A. H. L. Fizeau (1819–96), some six years later, showed that observations on the periodic variations of the precise positions of the spectral lines in the spectra of 'double stars' (the components of which revolve in orbits around their common centre of mass) were adequately explained on this basis. Other astronomical evidence of a similar nature soon accumulated, and the optical Doppler effect was added to

Bradley's 'aberration of light' as affording strong confirmation for the 'absolutist' view that we are discussing.

Naturally, with the luminiferous ether regarded as providing a standard of absolute rest, inquiry turned to the possibility of determining, by optical means, the 'true' velocity of the earth through the ether. The velocity of the earth in its orbit around the sun was known from astronomical observation, but, at the best, the velocity of the sun with respect to the nearer stars was known only approximately and on the basis of 'statistical' evidence. To determine the mean velocity of the earth through the ether, in a single series of experiments, would clinch the matter once and for all.

The first experiments designed for this purpose were carried out by A. A. Michelson (1852–1931) and E. W. Morley (1838–1923) between 1881 and 1887. In this case a second-order effect was involved. The time of back-and-forth travel of a light signal, over a fixed distance, is greater in the ratio $1:(1 - v^2/c^2)$ when the whole system of source, adjacent observer and distant mirror (for the reflection of the light) is moving through the ether with velocity v (in the direction of propagation) than when the system is 'absolutely' at rest. Michelson and Morley divided a beam of monochromatic light into two by means of a 'half-silvered' mirror. Directing the two beams so obtained along paths at right angles to one another, they arranged plane mirrors to reflect these beams back along their original paths. At the half-silvered mirror the beams were again divided, the reflected portion of the one becoming superposed on the transmitted portion of the other, as indicated (purely schematically) in Figure 28. With suitable adjustment of the apparatus, these superposed beams gave rise to interference fringes in the focal plane of the viewing telescope. The whole arrangement of light source, mirrors and telescope was freely supported on floats on a large trough of mercury. It was slowly rotated about a vertical axis, the fringe system being observed continuously. Since any change in the difference of the back-and-forth transit times for the two paths of the initially divided beam would show up as a displacement of the fringes, it appeared inconceivable, on the basis of the stagnant-ether view, that there should be no observable fringe shift during a complete rotation of the apparatus. The sensitivity had been reliably estimated as adequate for the detection of a velocity relative to the ether of less than one fifth of the orbital velocity of the earth around the sun. In spite of their confident anticipation, Michelson and Morley had to report that no fringe displacement was observed – at whatever time of year the experiment was repeated.

Before long other (second-order) effects, less obvious, perhaps, in their direct relevance to the problem, were found to imply the same conclusion – and theoretical physicists, for the first time (though not for the last), began, half in sheer frustration, half playfully, to speak of a 'conspiracy' of nature, a conspiracy, in this case, to hide from the inquiring observer all evidence of his 'real' velocity in absolute space. Different theorists reacted differently to this growing frustration. G. F. Fitzgerald (1851–1901) made the purely *ad hoc* suggestion, in 1893, that bodies in motion through the ether suffer deformation in the direction of their motion, lengths in this direction being reduced to $(1 - v^2/c^2)^{\frac{1}{2}}$ of their values when at rest. If that were so, then the light in the experiment of Michelson and Morley would not be travelling back and forth 'over a fixed distance', as naïvely supposed: this distance would vary as the apparatus rotated, and the null result would be understandable.

H. A. Lorentz (1853–1928), two years later, took a wider view. Maxwell's calculation of the speed of propagation of electromagnetic waves in 'free space' involved the solution of the differential equations ('Maxwell's equations') connecting the time and space variations of the electromagnetic field quantities. These equations, so Lorentz supposed – in conformity with the other leading theorists of the day – were 'really true' only in relation to the absolute space of the luminiferous ether (and absolute time). It seemed that they were 'apparently true' in relation to other spaces in uniform rectilinear motion through the ether (the experiment of Michelson and Morley might be understood as implying that the measured speed of light is the same in all such spaces). Lorentz set himself the problem of devising a set of transformation rules, to replace the equations of transformation characteristic of Newtonian relativity (see equations **8.1** and p. 166), which would ensure this result (which would make Maxwell's equations 'invariant with respect to the transformation'). By 1902 he was able to show that the following equations fulfilled this requirement completely:

$$x' = \gamma(x - vt),$$
$$y' = y,$$
$$z' = z, \hspace{4cm} \textbf{8.3}$$
$$t' = \gamma\left(t - \frac{vx}{c^2}\right),$$
$$\text{if} \quad \gamma = \left(1 - \frac{v^2}{c^2}\right)^{-\frac{1}{2}}.$$

In equations **8.3**, c represents the measure of the velocity of light in free space, as before – and the other symbols have precisely the meanings which we attached to them previously, in formulating equations **8.1**. Lorentz gave a full exposition of his 'theory' in a long paper published in 1904.

We may provide an example of equations **8.3** in use by showing how they include the 'Fitzgerald contraction' (see above) as a particular result. Let us consider a measuring rod at rest in the ether frame, lying along the x-axis. Let there be a rod-like body, lying along the x'-axis of the moving frame, at rest in that frame, with its ends at the points $(x_1', 0, 0)(x_2', 0, 0)$. Measurement of the 'true' length of the body (according to the absolutist point of view) consists in the observer in the moving frame recording the readings, x_1 and x_2, on the fixed measuring rod which, at an arbitrary instant of time (say t_0' according to his reckoning), correspond to x_1' and x_2', the end-points of the body as recorded in the moving frame. We conclude, then, that the pairs of specifications $(x_1, 0, 0, t_1)$ and $(x_1', 0, 0, t_0')$, and $(x_2, 0, 0, t_2)$ and $(x_2', 0, 0, t_0')$ represent, in the space-and-time coordinates of the two frames concerned, the two

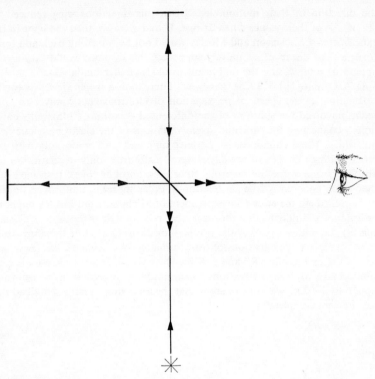

Figure 28

'events' involved in the measurement. These pairs of specifications being related by the first and last of equations **8.3**, we have

$$x_2' - x_1' = \gamma\{(x_2 - x_1 - v(t_2 - t_1)\}. \tag{8.4}$$

$$t_0' - t_0' = \gamma\left[t_2 - t_1 - \frac{v}{c^2}\left(x_2 - x_1\right)\right]. \tag{8.5}$$

From equation **8.5** we obtain

$$t_2 - t_1 = \frac{v}{c^2}\left(x_2 - x_1\right),$$

noting, in passing, that an observer in the ether frame would not regard the 'measuring events' as simultaneous, and, on substitution in equation **8.4**,

$$x_2' - x_1' = \gamma(x_2 - x_1)\left(1 - \frac{v^2}{c^2}\right),$$

$$\text{or} \quad x_2' - x_1' = (x_2 - x_1)\left(1 - \frac{v^2}{c^2}\right)^{\frac{1}{2}}. \tag{8.6}$$

According to equation **8.6**, $x'_2 - x'_1$, the apparent length of the body in the moving frame, is less than its true length, $x_2 - x_1$, by the factor $(1 - v^2/c^2)^{\frac{1}{2}}$, as Fitzgerald suggested. It is necessary only to issue a word of warning here: our treatment of the problem in terms of a 'true' and an 'apparent' length is essentially a 'nineteenth-century' treatment – but that is all we are involved in at the present time.

As a second example of the use of equations **8.3** let us consider how velocities are related in the two frames, if these transformation rules are accepted – and, in particular, what may be said about the light velocity c. Purely formally, on the basis of the equations as given,

$$x'_2 - x'_1 = \gamma\{x_2 - x_1 - v(t_2 - t_1)\},$$
$$y'_2 - y'_1 = y_2 - y_1,$$
$$z'_2 - z'_1 = z_2 - z_1,$$
$$t' - t'_1 = \gamma\left[t_2 - t_1 - \frac{v}{c^2}\left(x_2 - x_1\right)\right].$$

Using the last of these results together with each of the others in turn, we have

$$\frac{x'_2 - x'_1}{t'_2 - t'_1} = \frac{x_2 - x_1 - v(t_2 - t_1)}{t_2 - t_1 - (v/c^2)(x_2 - x_1)},$$

$$\frac{y'_2 - y'_1}{t'_2 - t'_1} = \frac{y_2 - y_1}{\gamma\{t_2 - t_1 - (v/c^2)(x_2 - x_1)\}},$$

$$\frac{z'_2 - z'_1}{t'_2 - t'_1} = \frac{z_2 - z_1}{\gamma\{t_2 - t_1 - (v/c^2)(x_2 - x_1)\}}.$$

Proceeding to the limit, we finally obtain

$$\dot{x}' = \frac{\dot{x} - v}{1 - v\dot{x}/c^2},$$

$$\dot{y}' = \frac{\dot{y}}{\gamma(1 - v\dot{x}/c)}, \qquad\qquad\qquad **8.7**$$

$$\dot{z}' = \frac{\dot{z}}{\gamma(1 - v\dot{x}/c^2)}.$$

Equations **8.7** are the transformation equations for velocities which we set out to derive, γ having the same meaning as before.

Suppose, now, that a 'representative particle' is moving in the ether frame with velocity c. If the particle is moving along the x-axis with this velocity, then $\dot{x} = c$, $\dot{y} = 0$, $\dot{z} = 0$, and equations **8.7** yield $\dot{x}' = c$, $\dot{y}' = 0$, $\dot{z}' = 0$. If it is moving along the y-axis, $\dot{x} = 0$, $\dot{y} = c$, $\dot{z} = 0$, and we have

$$\dot{x}' = -v, \; \dot{y}' = c(1 - v^2/c^2)^{\frac{1}{2}}, \; \dot{z}' = 0$$

giving $\dot{x}'^2 + \dot{y}'^2 + \dot{z}'^2 = c^2$ – and similarly, if the particle is moving along the

z-axis. In all three cases the measure of the velocity of the particle in the moving frame is c. For an arbitrary direction of motion in the ether frame, the same result emerges, though the relevant algebra is more complicated. The result is no more than we should have anticipated: we started (p. 171) from the assertion that the Lorentz transformation leaves Maxwell's equations unchanged in form when expressed in relation to the moving frame. If that assertion is true (and we have now gone some way towards verifying it), then of necessity a spherical light wave in the fixed ether appears as a spherical wave propagated with the same speed when viewed by an observer in the moving frame. Looking back to equations 8.3, we merely have one further remark to make. If $v > c$, obviously γ is imaginary and the equations become physically meaningless. In this respect, at least, the light velocity c is a limiting velocity of relative motion of an observer with respect to the Maxwellian ether; for velocities greater than this the formalism of the nineteenth-century physicists was unable to provide any statement of significance.

8.4 Relativity according to Einstein

If we multiply both members of the first of equations **8.3** by v/c^2, then, by combination with the last of these equations, after slight rearrangement, we obtain

$$t = \gamma\left(t' + \frac{vx'}{c^2}\right). \qquad \textbf{8.8}$$

Similarly, if we multiply both members of the last of equations **8.3** by v, by combination with the first of the equations, we obtain

$$x = \gamma(x' + vt'). \qquad \textbf{8.9}$$

It will be realized that equations **8.8** and **8.9**, taken together with the second and third of equations **8.3**, exhibit the transformation rules of Lorentz in reverse sense: the space and time coordinates relative to the ether frame are now given explicitly in terms of the coordinates relative to the moving frame – and $-v$, the effective velocity of the (fixed) ether frame with respect to the other (see the first of equations **8.7**, with $\dot{x} = 0$). On this basis we recognize at once, that the Lorentz transformation is symmetrical as between the two reference frames concerned. There is absolutely nothing in the mathematical formalism representing the transformation to suggest that the one frame is 'more absolute' than the other, or that it is in any other way unique. This consideration leads us directly to the relativity hypothesis of Einstein.

Albert Einstein (1879–1955) repudiated the stagnant-ether view from the outset. Taking over the concept of the inertial frame from Newtonian relativity, he asserted instead that not only are the laws of motion valid for all inertial frames (p. 165), but so are the laws of electromagnetism – indeed, so are all the laws of physics which are soundly based on observation. To be precise, in 1905, Einstein put forward the speculative hypothesis that there is an infinite set of reference frames, any one of which is as 'fundamental' as any other, such that any empirical law of observational physics may be given mathematical expression in identical terms in relation to any frame of the set. For any additional reference frame to be included in Einstein's infinite set, all that was required was that the frame should be in uniform rectilinear motion relative to every other frame included in the set.

We have used the phrase 'all the laws of physics' in stating Einstein's speculative view. Obviously, this phrase has much too wide a reference to provide the basis for any precise development of theory. On the other hand, a very considerable development may be made on the basis of no more than two particular assumptions, as Einstein himself showed. In order to carry out this development it is not necessary to postulate either that all the laws of motion, or that all the laws of electromagnetism, be valid for all the frames of the set – only that Newton's first law (the law of 'inertia') is so valid, and that the free-space speed of light is the same for all the frames concerned. These, indeed, were the two basic postulates of Einstein's 'special theory' of 1905.

The law of inertia states that an 'isolated' particle moves with constant velocity indefinitely. Einstein's first postulate, therefore, was that a constant velocity in one frame appears as a constant velocity (in general of different magnitude and direction) in any other frame of the set. This postulate being accepted, the second merely relates to a special case. It adds the assertion that when the constant velocity in question is the free-space velocity of light, the magnitude of this velocity is the same for observers in all the frames concerned.

Now, we have already agreed (p. 174) that in respect of the Lorentz transformation (which was devised so that Maxwell's equations should be invariant in terms of it), the particular result of constant light velocity is a built-in requirement. In the matter of Einstein's first postulate, also, the same is true – albeit incidentally, rather than by primary design. This conclusion emerges directly from a consideration of equations 8.7: if the velocity components $\dot{x}$, $\dot{y}$, $\dot{z}$ remain constant indefinitely, then the corresponding velocity components $\dot{x}'$, $\dot{y}'$, $\dot{z}'$ remain constant, also (neither the space coordinates (x, y, z), nor the time t, are involved in these equations). Not surprisingly, therefore, in view of these results, when Einstein examined the logical consequences of the adoption of his two postulates, he was led back to the formalism of the Lorentz transformation. He was able to show that acceptance of these postulates carried with it the necessity that the transformation rules, as between any two reference frames of his infinite set, should be precisely as specified by the equations of Lorentz. In a sense, the matter had come full circle – in relation to the Michelson–Morley experiment, for example, the formal basis of interpretation was the same as before – but the philosophical standpoint had changed completely. Lorentz had regarded his transformation equations as relating observations (made for reasons of convenience) in a 'moving' frame to those in a 'privileged' frame fixed in the luminiferous ether: Einstein had abandoned the ether concept altogether, and all notions of privilege as irrelevant in this context – for him equations **8.3** had reference to any two frames of an infinite set of frames, symmetrically, none of these frames being any more fundamental than any other in relation to the real world.

We have said that Einstein asserted that all the laws of motion and all the laws of electromagnetism should be capable of mathematical formulation in all inertial frames of reference – each law being expressible in identical terms to whichever such frame it is referred. On the other hand, a much weaker requirement than this – namely the acceptance of the two basic postulates of the special theory (and the requirement of entire symmetry as between any two inertial frames) – is sufficient to lead to unique transformation rules, as we have just explained. Expressed in this unspecific form, this 'favourable' result need cause no surprise. Expressed more precisely in the statement that the necessary transformation rules are those of Lorentz, it carries a wider implication which reveals an uncovenanted 'success'. If the laws of electromagnetism 'really' have the structure that Maxwell's equations imply, then the result regarding the transformation rules ensures the validity of these equations in relation to all inertial frames – and that part of Einstein's general assertion which refers to the laws of electromagnetism is fully substantiated. Indeed, it is a logical consequence of the 'weaker requirement' of the basic postulates of the special theory (provided that Maxwell's equations are 'really true').

The situation is very different in relation to 'all the laws of motion'. Einstein's basic postulates include only Newton's first law (throughout this section we are using the term 'inertial frame' specifically in relation to this law: an inertial frame is simply one in which an 'isolated' particle moves with constant velocity indefinitely). We have to inquire what the position is in relation to the other laws of motion. In Newtonian relativity, acceleration (but not velocity) appeared as an invariant quantity as between two frames of reference in uniform rectilinear relative motion (p. 162). In the relativity of Einstein, neither velocity nor acceleration is invariant in this sense. Equations **8.7** state the transformation rules in respect of the velocity components of a representative particle according to Einstein's assumptions; it is only a matter of simple algebra to show that the transformation rules in respect of the components of acceleration of such a particle are equally complex. If, then, acceleration is not invariant in Einstein's relativity, the Newtonian definition of force is unworkable – at least on the basis of invariable mass. So much would have become apparent to anyone, whether he approached the matter from the philosophical standpoint of Einstein or not, who attempted to reconcile the equations of the Lorentz transformation with the equations of Newtonian dynamics at the beginning of this century.

It is a major part of Einstein's achievement that he did not, in 1905, accept this apparently negative result as the end of the story. Indeed, it is probable that he was aware of it from the beginning, and that we should have been more accurate historically (though we should thereby have sacrificed a profitable avenue of presentation) if we had taken count of this feature of the situation at the outset, and stated Einstein's conviction somewhat differently. 'The laws of motion, suitably modified, and all the laws of electromagnetism, should be expressible each in identical terms in whatever inertial frame is used for the purpose,' would have been a more appropriate expression of his revolutionary point of view. Our present task, then, is to consider the modifications which Einstein found it necessary to introduce into the Newtonian laws in order to substantiate his conviction. Obviously, we should not be accepting this task, if he had not been proved successful in the outcome.

It is a fact of history that in 1905 there was already almost incontrovertible evidence for the non-constancy of inertial mass – at least so far as the mass of the electron was concerned. W. Kaufmann had shown, in 1901, that the value obtained for the specific charge of the β-particles (negative electrons) from radium products, as a result of electric and magnetic deflection experiments, decreased regularly as the speed of the particles increased. The *specific charge* is the quantity e/m, e being the charge and m the inertial mass of the particles, and it was natural to assume that the variability of this quantity arose from variations in mass rather than from variations in charge.

Any reader will rightly be suspicious of statements which are mixtures of objective fact and speculative interpretation, and we should pause to justify the interpretative content of our last paragraph before we proceed. We have said that it was natural to regard Kaufmann's results as almost incontrovertible evidence for the non-constancy of the inertial mass of the electron, in 1901. To provide the reason for this statement, it is necessary to go back another twenty years in time. According to Newtonian mechanics work has to be done to set a body in motion with constant velocity. The measure of this work is proportional to the product of the inertial mass of the body and the square of the final velocity of motion. In 1881, J. J. Thomson (1856–1940) drew attention to the fact that, according to the laws of electromagnetism, this work is greater when the body is charged than when it is uncharged, the excess work, also, being proportional to the square of the final velocity of motion when that velocity is small compared with the velocity of light. It is as if the charge on the body increases the inertial mass by a small contribution 'of electromagnetic origin'. For ordinary macroscopic bodies this contribution is relatively insignificant (for a water droplet of one-millimetre radius charged to a potential of 100 V it represents about three parts in 10^{21} of the normal mass of the droplet), but for particles of atomic dimensions the situation might be otherwise. In 1897, largely through the experimental investigations of Thomson himself, the negative electron was discovered as a common constituent of the atoms of all chemical substances, carrying the natural unit of charge and having an inertial mass smaller than that of the simplest atom by a factor of a thousand or more. The question immediately arose whether in this case the inertial mass could be entirely of electromagnetic origin (if it were, the radius of the electron would be of the order of 2×10^{-15} m, an entirely acceptable value). If this were the case, Thomson's theoretical investigation of 1881 indicated that the inertial mass of the electron should vary significantly with its speed (the work required to set a moving charge in motion increases, according to the laws of electromagnetism, more rapidly than the square of the final velocity, when that velocity becomes comparable with the velocity of propagation of electromagnetic waves in the surrounding medium).

So we have provided the background to our previously unjustified assertion regarding the view which was held, by theoretical physicists generally, concerning the experimental results of Kaufmann in 1901. Although these results ran contrary to the Newtonian view of the invariability of inertial mass, they were not unexpected (as the results of Michelson and Morley were, in relation to the stagnant-ether view current at the time). Indeed, Kaufmann's results were regarded with general satisfaction, as advancing the view that the mass of the electron – the first identified constituent of the atom as an electrical structure – might prove, in the end, to be an attribute of its intrinsic electrification and nothing more.

The reason for our digression, from Einstein back to Kaufmann and Thomson, will now be obvious. We have said that the Newtonian definition of force is unworkable in Einstein's relativity, 'at least on the basis of invariable mass' (p. 176) – and we have also implied that Einstein was able to overcome this difficulty by suitably modifying the Newtonian laws. We have set ourselves the task of considering the modifications that were necessary to this end. From what we have now said concerning Kaufmann's experiments (and their interpretation), it will at least be clear that when Einstein was developing his special theory in 1905, if he had not provided, within the scope of the theory, an explanation of the variability of mass, he would have failed in his aim.

In section 6.9 we drew attention to the fact that 'the Newtonian scheme ... is characterized by a basic simplicity of a very fundamental kind.... The concepts of mass and (linear) momentum which Newton introduced ... are of such a character that, in their global reference, they identify quantitative aspects of permanence in the real world.' We were dealing in that section with the conservation laws of Newtonian dynamics – and we saw these laws as the foundation of its success.

Essentially, in 1905, Einstein set himself the problem of so modifying the Newtonian definition of mass that the conservation laws of mass and momentum should be retained for all inertial frames (between any two of which the Lorentz transformation provided the appropriate rule for relating the space and time coordinates of observed events). Accepting the necessity that mass should vary with speed (as observed in any particular frame), Einstein retained the Newtonian definition of momentum – 'the measure of it, arising from its velocity and its mass conjointly' – and was able to show that a unique law of speed dependence was thereby specified. In any inertial frame, an observer at rest with respect to the frame would assign to a particle moving with velocity u in that frame a mass which varied with u as $(1 - u^2/c^2)^{-\frac{1}{2}}$. According to this specification, the mass of any particle would tend to infinity as the speed of the particle approached the free-space speed of light. Obviously, on this basis, no particle could ever be observed, in an inertial frame, travelling with a speed greater than this.

8.5   Einstein's relativity: an alternative approach

We have not, in the last section, given any formal derivation of Einstein's various results, and we shall not do so now. We shall, however, consider the matter of mass variation in some detail from another point of view from which interesting results emerge. In Einstein's derivation of the mass-variation equation, the Lorentz transformation was assumed. As a consequence, the free-space speed of light appeared as the limiting speed of observable motion of a material particle. In a sense, we shall reverse the process: we shall assume that there is a limiting speed of observable particle motion (not necessarily the speed of light), and in the course of our argument (admittedly an incomplete and non-rigorous argument, relying for its 'validity' on a full treatment by R. Weinstock, 1965) we shall derive a velocity-transformation rule which is formally identical with that of Lorentz. In the end, we shall obtain the Einstein mass relation, once we have identified our limiting particle speed with c, the free-space speed of propagation of light.

Let us suppose, then, that we have two inertial frames S and S' relatively orientated as in Figure 26(p. 162), S' moving relatively to S with velocity v along OX. Consider the collision of two particles in motion (as seen by an observer in either frame) in OX(OX'). Essentially we are considering a one-dimensional situation. According to an observer in S, the masses of the particles before collision are m_1 and μ_1, and the measures of the corresponding momenta are p_1 and π_1. After the collision, the masses are m_2 and μ_2, and the measures of the momenta p_2 and π_2. Let us employ dashed symbols m_1', p_1', ... to represent these same quantities, as measured by an observer in S'. Both observers, so we assume, find the conservation laws to be exemplified by their measurements. Formally, we have, for the observer in S,

$$m_1 + \mu_1 = m_2 + \mu_2,$$
$$p_1 + \pi_1 = p_2 + \pi_2; \qquad\qquad \textbf{8.10}$$

and, for the observer in S',

$$m_1' + \mu_1' = m_2' + \mu_2',$$
$$p_1' + \pi_1' = p_2' + \pi_2'. \qquad\qquad \textbf{8.11}$$

Our task is to devise transformation rules for the (related) quantities mass and momentum in terms of which equations **8.11** may be satisfied simultaneously with equations **8.10**.

Obviously, a sufficient condition for the simultaneous validity of these two pairs of equations (we do not attempt to show that it is a necessary condition) is that mass in S' shall be related linearly to mass and momentum in S – and

179   Relativity According to Einstein

similarly in respect of momentum in S'. If A, B, C and D are constants for a given pair of inertial frames in specified relative motion, and if we write

$$m' = Am + Bp,$$
$$p' = Cp + Dm; \qquad\qquad \textbf{8.12}$$

clearly
$$m_1' + \mu_1' = A(m_1 + \mu_1) + B(p_1 + \pi_1),$$
$$m_2' + \mu_2' = A(m_2 + \mu_2) + B(p_2 + \pi_2),$$

and, on this basis, if equations **8.10** are valid, then the first of equations **8.11** is valid, also. A similar analysis establishes the validity of the second of equations **8.11**, under the same assumptions.

We have now found transformation rules (equations **8.12**) for mass and momentum which fulfil our basic requirements: let us look at them in more detail. We naturally assume that all four constants, A, B, C and D, involve v the measure of the velocity of relative motion of S' and S. Clearly, A and C are non-dimensional constants, and in the limit when v is zero each must take the value unity. B and D, on the other hand, are dimensional constants, and when v is zero their measures must be zero, also. In order to make A and C non-dimensional, obviously we must introduce a characteristic speed which is extraneous to our particular case, and which we assume to be a fundamental constant – at least in the context of equations **8.12**. Let us denote the measure of this velocity by k. Then our conclusions, at this stage, may be represented formally as follows

$$\left[A\!\left(\frac{v}{k}\right) \right]_0 = \left[C\!\left(\frac{v}{k}\right) \right]_0 = 1,$$
$$\{B(v)\}_0 = \{D(v)\}_0 = 0. \qquad\qquad \textbf{8.13}$$

We now introduce the principle of the equivalence of inertial frames. On that basis, in conformity with equations **8.12**, we must have

$$m = A'm' + B'p',$$
$$p = C'p' + D'm'. \qquad\qquad \textbf{8.14}$$

Here, for brevity, we have written A' for $A(-v/k)$, B' for $B(-v)$, and correspondingly for C' and D'. Alternatively, by direct solution of equations **8.12**, we obtain

$$m(AC - BD) = Cm' - Bp',$$
$$p(AC - BD) = Ap' - Dm'. \qquad\qquad \textbf{8.15}$$

Comparison of equations **8.15** with equations **8.14** yields the obvious solution

$$B' \equiv B(-v) = -B(v),$$
$$D' \equiv D(-v) = -D(v), \qquad\qquad \textbf{8.16}$$

and the subsidiary results

$$\left. \begin{aligned} AC - BD &= 1, \\ A' &= C, \\ C' &= A. \end{aligned} \right\} \qquad\qquad \textbf{8.17}$$

According to equations **8.16**, B and D change sign with v. Now B is of dimension -1, and D is of dimension $+1$, in velocity (equations **8.12**), and each tends to zero with v (equations **8.13**). We satisfy these requirements, again introducing our characteristic speed k, writing

$$B(v) = E\left(\frac{v^2}{k^2}\right)\frac{v}{k^2},$$

$$D(v) = F\left(\frac{v^2}{k^2}\right)v.$$

8.18

In relation to equations **8.18**, we merely note, here, that the non-dimensional quantities E and F involve the even powers of v/k, exclusively.

Let us now write $p = mu$, $p' = m'u'$, u and u' being velocities along $OX(OX')$, as determined by the observers in S and S', respectively, and let us rewrite equations **8.12**, incorporating the results of equations **8.18**. We have

$$m' = Am + E\frac{muv}{k^2},$$

$$m'u' = Cmu + Fmv.$$

From these two equations, we obtain

$$\frac{m'}{m} = A + E\frac{uv}{k^2} = C\frac{u}{u'} + F\frac{v}{u'}.$$

8.19

It is a simple interpretation of the meaning of the phrase 'relative velocity', in relation to two inertial frames which are equivalent, that when $u' = 0$, $u = v$, and when $u = 0$, $u' = -v$. Substituting these pairs of values, in turn, in equations **8.19**, we have

$$C + F = 0,$$
$$A = -F.$$

Thus $A = C = -F$. **8.20**

The first of equations **8.20** taken together with the second of equations **8.17** implies $A' = A$, that is

$$A\left(\frac{-v}{k}\right) = A\left(\frac{v}{k}\right);$$

8.21

then from the second and third of equations **8.17**, $C' = C$. Equation **8.21** requires that only the even powers of v/k are involved in A: we have already concluded (equations **8.18**) that the same is true of F, also.

Let us now substitute from equations **8.18** and equations **8.20** in the first of equations **8.17**. We have

$$A^2 + AE\frac{v^2}{k^2} = 1.$$

8.22

Substituting from equations **8.20** in equations **8.19**, we likewise obtain

$$A + E\frac{uv}{k^2} = A\frac{(u-v)}{u'},$$

or $\quad u'\left(1 + \frac{E}{A}\frac{uv}{k^2}\right) = u - v.$ **8.23**

A and E, as we have concluded, are functions of v^2/k^2 only. Equation **8.22**, therefore, is a functional relation in this variable, exclusively. Equation **8.23**, on the other hand, involves u and u' – it is, indeed, a formal statement, not yet entirely explicit, of the rule of velocity transformation as between inertial frames. Although we have the two equations, we cannot solve for A and E separately (as functions of v^2/k^2); we need to incorporate some other assumption.

We have introduced k as a 'characteristic speed' which by implication is a universal constant. Let us now assume that this speed is the 'limiting speed of observable particle motion' to which we referred at the beginning of this section. By this identification we imply that there is a 'limitingly possible' state of motion of a particle which observers in all inertial frames would recognize as being executed with the same speed k. Such a case would be represented by $u = k$, $u' = k$, in equation **8.23**. In relation to such a case, therefore,

$$k\left(1 + \frac{E}{A}\frac{v}{k}\right) = k - v,$$

or $\quad E = -A.$ **8.24**

Substituting now, from equation **8.24**, in equations **8.22** and **8.23** in turn, we have

$$A = \left(1 - \frac{v^2}{k^2}\right)^{-\frac{1}{2}},$$ **8.25**

$$u' = \frac{u-v}{1 - uv/k^2};$$ **8.26**

and substituting from equations **8.24** and **8.25** in the first of equations **8.19**,

$$\frac{m'}{m} = \left(1 - \frac{v^2}{k^2}\right)^{-\frac{1}{2}}\left(1 - \frac{uv}{k^2}\right).$$ **8.27**

According to the principle of the equivalence of inertial frames, we must suppose that the law of velocity dependence of mass is the same, whether it is derived from measurements in S or S'. Furthermore, we naturally assume that, in a given frame, mass is independent of the direction of motion of a particle, its speed being given. In our present context, therefore, we require

$$\frac{m'}{m} = \frac{f(u'^2/k^2)}{f(u^2/k^2)}.$$ 8.28

There is no obviously simple way of solving the simultaneous equations 8.27 and 8.28, but fortunately a first guess (on almost anyone's part!)* turns out to be successful. Taking account of equation 8.26, it is not difficult to establish the identity

$$\left(1 - \frac{u'^2}{k^2}\right)\left(1 - \frac{uv}{k^2}\right)^2 \equiv \left(1 - \frac{u^2}{k^2}\right)\left(1 - \frac{v^2}{k^2}\right);$$ 8.29

the left-hand member may be written

$$\left(1 - \frac{uv}{k^2}\right)^2 - \left(\frac{u - v}{k}\right)^2,$$

which is $\quad 1 - \dfrac{u^2}{k^2} - \dfrac{v^2}{k^2} + \dfrac{u^2 v^2}{k^4},$

which is identical with the right-hand member, as was to be shown. On the basis of identity 8.29, therefore, and equation 8.27 we have, finally,

$$\frac{m'}{m} = \left(\frac{1 - u^2/k^2}{1 - u'^2/k^2}\right)^{\frac{1}{2}},$$

or $\quad m\left(1 - \dfrac{u^2}{k^2}\right)^{\frac{1}{2}} = m'\left(1 - \dfrac{u'^2}{k^2}\right)^{\frac{1}{2}} = m_0.$ 8.30

In equations 8.30, the quantity m_0 is obviously the mass which would be assigned by an observer, whether in S or S', to a particle moving with vanishingly small velocity in his own frame. This quantity is referred to as the 'rest mass' of the particle – and the equations show that, at any other speed, in any inertial frame, the mass of the particle will be estimated as greater than m_0 by the factor $(1 - u^2/k^2)^{-\frac{1}{2}}$, u being the speed of the particle as observed in that frame. In equations 8.30, if $k = c$, we have the mass–speed relationship of Einstein (p. 178) – and, on the basis of the same identification, we have, in addition, the velocity transformation rule of Lorentz, as it applies to our particular case, in equation 8.26 (see equations 8.7).

8.6   Recapitulation and discussion

Twentieth-century 'special' relativity rests on the assumption of the fundamental equivalence of inertial reference frames in uniform rectilinear motion, one with respect to another. In 1905, Einstein showed that the transformation

---

* Anyone who noticed that when $u = v$, $u' = 0$, $m = m'(1 - v^2/c^2)^{-\frac{1}{2}}$ would almost certainly make this his first trial solution of the equations concerned.

rules relating the measurements which observers in two such frames would make, on a series of events open to their common observation, could be deduced unambiguously on the basis of two simple postulates. These were that the law of inertia is valid in each frame, and that the free-space speed of light has the same value in each. The choice of these postulates – one relating to a law of mechanics, and the other to a phenomenon in electromagnetism – was symbolic of Einstein's underlying philosophy, that the whole of physics, and these two major branches in particular, must disclose itself in the same way to observers in all the frames of the 'inertial' set, the same mathematical laws emerging from the individual measurements of each observer.

In section 8.4 we stated Einstein's conclusions without attempting their logical derivation. On the basis of the two postulates that we have quoted, Einstein showed that the transformation rules for space and time coordinates to which he was led were identical with those which (motivated by an altogether different philosophy) Lorentz had devised for the purpose of rendering the formulation of the electromagnetic theory of Maxwell invariant as between inertial frames. Adding to these postulates those of the conservation of mass and momentum in particle collisions in all inertial frames, Einstein further showed that particle mass must be regarded as speed dependent, the necessary variation law being of the form $m = m_0(1 - u^2/c^2)^{-\frac{1}{2}}$, as we have already stated. Basically, we may say that the primary experimental facts justifying these postulates are the results of the Michelson–Morley experiment (interpreted as indicating that the measured speed of light is independent of the (inertial) frame in relation to which it is measured), and those of ordinary, 'low-velocity' collision experiments in the tradition of Newton and Wren. The results of Kaufmann, on the apparent variation of the inertial mass of negative electrons with the speed of those particles, may be regarded as providing evidence of a confirmatory nature: they can hardly be regarded as providing any essential primary data.

In section 8.5 we developed in some detail (but still with restricted generality) an alternative approach. We accepted the fundamental equivalence of inertial frames, and the conservation of mass and momentum in particle collisions in all such frames. But we did not use Einstein's first postulate explicitly, and we replaced his second postulate by what is at first sight a different one. The second postulate of Einstein states, in effect, that there is a characteristic speed c, which is the same for all inertial frames, which is the free-space speed of light as measured by an observer in any such frame. We replaced this postulate with the axiomatic statement that there is a characteristic speed k, which is the same for all frames, such that if a particle is observed to have a velocity of this magnitude with respect to any inertial frame it will be observed to have a velocity of the same magnitude with respect to any other such frame. If the results of our particular investigation can be generalized – as, in fact, they can – then we may conclude that this approach leads to conclusions that are formally identical to those of Einstein: the velocity-transformation rules are the Lorentz rules (and the coordinate transformation rules must be those of Lorentz, also), and mass is shown to vary with observed speed in the same way as before. The only difference is that our characteristic speed k has replaced Einstein's light speed c in all the relevant equations.

It appears, then, that in respect of form, at least, all the transformation rules of special relativity can be derived either on the basis of Einstein's postulates, or on the basis of the alternative postulates of section 8.5. The question arises: what is the implication of this duality? If we adopt the alternative approach, we are dealing entirely in mechanical concepts from the outset – and there is no reason in principle why we should not determine the characteristic speed k which this approach identifies, from purely mechanical experiments. The fact that most of our information concerning the variation of mass with speed comes indirectly, or, if directly, then from deflection experiments involving charged particles, need not complicate the issue. If we adopt our alternative approach, we may claim, without fear of contradiction, that $k = 2 \cdot 998 \times 10^8$ m s^{-1} is a straightforward experimental result.

Continuing with this point of view, we note, as an entirely independent result of experiment carried out in a terrestrial laboratory, that

$$(\mu_0 \, \epsilon_0)^{-\frac{1}{2}} = 2 \cdot 998 \times 10^8 \text{ m s}^{-1}.$$

Here μ_0 and ϵ_0 are the permeability and permittivity 'of free space' – and $(\mu_0 \, \epsilon_0)^{-\frac{1}{2}}$ is the measure of the speed of propagation of electromagnetic waves according to Maxwell's equations. It appears coincidental that, within the limits of experimental uncertainty, $k = (\mu_0 \, \epsilon_0)^{-\frac{1}{2}}$ – but, if it were not so, Maxwell's equations would not be strictly invariant, and the apparent null result of the Michelson–Morley experiment might conceal a slight, but real, effect.

In contrast with the situation that we have just described, Einstein's derivation of the transformation rules, as we have seen, is based on a mixture of postulates, referring to mechanical and electromagnetic concepts in conjunction. From the outset, the characteristic speed which is involved is the experimentally determined free-space speed of light propagation: $(\mu_0 \, \epsilon_0)^{-\frac{1}{2}}$, according to the theory of Maxwell. From this point of view it is no coincidence that this speed is involved, also, in the mass–speed relation.

What are we to comment? No physicist would be so naïve as to believe that it is purely coincidental that $k = c$ (we are using the language of our alternative approach) – but neither would he be so ingenuous as to suppose that it is a matter of no interest that the Lorentz transformation rules (which render Maxwell's equations invariant when the characteristic speed involved has the 'correct' value) can be derived on the basis of purely mechanical experiments and their logical evaluation. We can say only one thing for certain: on either basis the law of mass variation is entirely independent of any particular assumption regarding the 'origin' of inertial mass. The speculations of Thomson (p. 177), and those who later extended his calculations, have no direct relevance to the perfectly general question with which we are here concerned.

185   Recapitulation and Discussion

8.7 Force, mass and energy

Originally, we set ourselves the task (p. 177) of considering the modifications which Einstein introduced into the Newtonian laws in order to justify his conviction that 'all the laws of physics which are soundly based on observation' (p. 174) are equally valid for observers in all inertial frames. So far, we have merely considered the modification which proved necessary in relation to the Newtonian principle of the invariability of mass. We have now to extend the scope of our inquiry to include the concepts of force and energy, also.

In the Newtonian scheme, the concept of force becomes significant once mass has been defined – in terms of the 'sudden' changes of velocity which take place in particle collisions, or, more generally, in terms of the ratios of instantaneous accelerations of particles interacting at a distance. This definition of mass leads to the principle of the conservation of momentum (as expressing a fact about the real world) and, when force has been suitably defined, to Newton's third law concerning the pairing of forces in action–reaction pairs.

In relativistic particle dynamics we retain the Newtonian definition of the (vector) force as the time rate of change of the (vector) momentum, but, because the (scalar) mass is now speed dependent, we can no longer equate the measure of the force to the product of the measures of mass (either rest mass or effective mass) and acceleration. In general, indeed, for a particle in accelerated motion, the directions of the unbalanced force and the instantaneous acceleration are not the same. In deriving the law of mass variation with speed, we have already accepted the principle of conservation of momentum in particle collisions. On the basis of the relativistic definition of force, clearly this is equivalent to establishing the equality of action and reaction in this restricted context. However a little consideration will show that, relativistically, Newton's third law is not of universal validity. The pairing of forces is a principle which is valid in the limit of 'point events', is approximated to with considerable accuracy in a 'local' domain, but is meaningless when 'distant' interactions are involved.

In order to justify the statement that we have just made, it is necessary only to look carefully at the assumptions which underlie the Newtonian formulation, and to confront these assumptions with the requirements of the Lorentz rules. According to the Newtonian philosophy, it is possible in principle, however large the system concerned, to identify the instantaneous accelerations of all the constituent particles of the system 'at an instant in time' – and it is implicitly assumed that two observers, even though they might be in uniform relative motion rectilinearly, would have a common appreciation of simultaneity, and a common ordering in time of all events within a common field of observation. According to the transformation rules of Lorentz, these assumptions are denied validity. We have already noted in passing (p. 172) the disagreement over the respective notions of simultaneity, as between observers in two inertial frames, which is inherent in the Lorentz rules. The result which we then obtained on the basis of equation **8.5** shows quite clearly that if A and B are two events which in one frame (S′) are regarded as simultaneous, then A may be judged to precede B, or to succeed it, by an observer in the other frame (S), depending on the conditions obtaining. Neither simultaneity, therefore, nor, in all circumstances, order-in-time is a validly tansferable concept as between observers in different inertial frames, according to the relativity of Einstein.

We reiterate: the whole notion of action at a distance loses its simplicity in twentieth-century relativity, and in consequence Newton's third law loses its central position as a simple statement concerning the ultimate nature of the real world. Always provided that actions are propagated (according to the older view) through empty space with the speed of light, no real difficulty arises – for the transformation rules ensure invariancy – but, when this is not the case, even 'special relativity' fails to provide a satisfactorily invariant description of the process. We shall refer to this possibility, again, briefly at the end of this section.

Just as, in special relativity, we take over the Newtonian definition of force (in one of its forms), so we take over the unique definitions of work and kinetic energy from Newtonian dynamics. We say that the work performed by a force is given by the scalar product of the force and the distance through which its point of application moves – and we assert that if this force is the total unbalanced force acting on a particle, then the kinetic energy of the particle is increased, by the action of the force, by an amount equal to the work done. Let us consider the one-dimensional case in which (in a particular inertial frame) a particle moving along the x-axis between the points $(x_1, 0, 0)$ and $(x_2, 0, 0)$ suffers an increase in kinetic energy from T_1 to T_2, the momentum of the particle increasing from p_1 to p_2 and the measure of its velocity from u_1 to u_2 in the time interval t_1 to t_2. If F represents the instantaneous value of the force acting on the particle, we have

$$T_2 - T_1 = \int_{x_1}^{x_2} F \, dx,$$

$$F = \frac{dp}{dt},$$

therefore $\quad T_2 - T_1 = \int_{t_1}^{t_2} \frac{dp}{dt} \frac{dx}{dt} \, dt,$

$$= \int_{p_1}^{p_2} u \, dp,$$

$$= [up]_1^2 - \int_{u_1}^{u_2} p \, du.$$

Now $p = \dfrac{m_0 u}{(1 - u^2/c^2)^{\frac{1}{2}}}$,

m_0 being the rest mass of the particle.

Thus
$$T_2 - T_1 = m_0 \left[\frac{u^2}{(1 - u^2/c^2)^{\frac{1}{2}}} + c^2 \left(1 - \frac{u^2}{c^2} \right)^{\frac{1}{2}} \right]_1^2 ,$$

$$= m_0 c^2 \left(\frac{1}{(1 - u_2^2/c^2)^{\frac{1}{2}}} - \frac{1}{(1 - u_1^2/c^2)^{\frac{1}{2}}} \right),$$

$$= (m_2 - m_1)c^2,$$

m_1 and m_2 being the values of m, the effective mass of the particle, when the measures of its velocity are u_1 and u_2, respectively. By definition, $m = m_0$ when $u = 0$. Thus we have, generally,

$$T = (m - m_0)c^2. \qquad\qquad\qquad 8.31$$

Assessing the significance of equation 8.31, we note, first of all, that it reduces to the Newtonian form (as it must, if it is to be acceptable) when $u \ll c$. Substituting for m, we have, in fact,

$$T = \frac{1}{2} m_0 u^2 \left(1 + \frac{3}{4} \frac{u^2}{c^2} + \dots \right).$$

More fundamentally, equation 8.31 suggests a general relationship between effective inertial mass and energy – with the 'universal constant', the characteristic speed c, the free-space speed of light, providing the 'conversion factor' c^2. Obviously, we cannot here follow through the argument establishing the generality of this relationship. Suffice to say that it emerges from the special theory with full generality: whether, in the formal dynamics of particles, we are concerned with potential energy or kinetic energy – or, in relation to 'real' atoms, with internal 'energy of excitation', or energy of rotation, or energy of translatory motion – whenever energy is any of these forms becomes associated with an identifiable entity, the effective mass of the entity is thereby increased by the corresponding amount. Denoting the increment of energy, of whatever form, by ΔW, and the increment of mass by Δm, we have $\Delta W = c^2 \, \Delta m$, in all cases. Over the last thirty years this relationship has been verified, to a high order of precision, by innumerable experiments in the field of nuclear transmutation: indirectly, these experiments may also be regarded as providing the most convincing evidence in support of the relativistic law of mass variation with speed, which, as our derivation of equation 8.31 indicates, is intimately involved in the result that we are discussing.

We revert finally to our recent statement that Einstein's special theory of relativity fails to provide a satisfactorily invariant description of any natural phenomenon which involves – or appears to involve – action at a distance, when that action is not propagated through empty space with the speed of light. Einstein recognized, from the outset, that the phenomenon of gravitational attraction is of this character. To his ear, the grand eloquence of Newton's universal law (p. 141) had an empty ring – and the problem of the apparent duality of the mass concept (p. 143) was an additional cause for concern. By 1915 he had so extended his theory that gravitational effects could be described in a mathematical language in which the concept of action at a distance was not involved – or the Newtonian notion of the pairing of forces. This extended theory has come to be known as Einstein's 'general theory' of relativity. Until very recently it had not been possible to submit to the test of laboratory experiment, in any convincing way, any of the predictions of this theory, though it had had its successes in one or two instances in the field of observational astronomy (see p. 161). For this reason, and more importantly because the theory is mathematical to a degree altogether beyond the scope which we can claim for our present account, we must, however, abandon any further attempt to characterize it in detail.

Further reading

A. B. Arons, *Development of the Concepts of Physics* (Chapter 36), Addison-Wesley, 1965.

C. W. Kilmister, *The Environment in Modern Physics* (Chapters 1–5), English Universities Press, 1965.

R. B. Lindsay, *Physical Mechanics* (3rd edn, Chapter 14), Van Nostrand, 1961.

W. C. Michels, M. Correll and A. L. Patterson, *Foundations of Physics* (Chapter 8), Van Nostrand, 1968.

R. Weinstock, 'New Approach to Special Relativity', *American Journal of Physics*, vol. 33 (1965), pp. 640–45.

Index